Sira Hentze

land
Flyvers
Platform
Pilen
Krydset
tionen
Fredens Have
Søminen

Supported by:

Danish Arts Foundation

dreyersfond

BECKETT·FONDEN

Danmarks Nationalbank's Anniversary Foundation

The Plum Foundation for Peace,
Organic Development and Sustainability

THE FLOATING COMMUNITY

Henrik Valeur with Francisca Calderon and Frédéric Duhau,
current and former inhabitants:
Andreas Wolf, Gabriel, Kami, Kasper Hedegreen, NN, Örn,
Sira Hentze, Stephen James Back and Storken.

The Architectural Publisher B
Copenhagen, Denmark
www.b-arki.dk
Layout: Gilbert Hansen
Copy-editor: Dan Marmorstein
Printing: Narayana Press
ISBN: 978-87-92700-36-0

Images by Henrik Valeur, his team and the participants, Creative Commons, Google Earth and other sources where mentioned.

The content is based on three years action research on site.

A HAVEN ON WATER

Havn] and the Pirate Harbor – is a spont
for people living on the fringes of socie
centrally in Copenhagen, right next to th
town, Christiania; an upscale residential
institutions on Holmen. For more than
small self-refurbished boats and small
around existing jetties, mooring poles
differs through the seasons, with as fe
and up to several dozens during warm
artists, poets and performers, lost so
modern society, with its fixed and uni
diversity, it is prone to both antag

he floating community – variously known
as the Harbor of Peace [In Danish: *Fredens*
eously evolved and self-organized haven
although it happens to be located very
vorld famous restaurant, Noma; the free-
ea; and a cluster of creative educational
lecade, people have been living here in
f-constructed houses on rafts clustered
d sunken boats. The number of people
as a handful or two during cold winters
mmers. This is a community of outsider
and dreamers, who don't fit into the
m way of living. Being a community of
stic disruption and mutual fertilization.

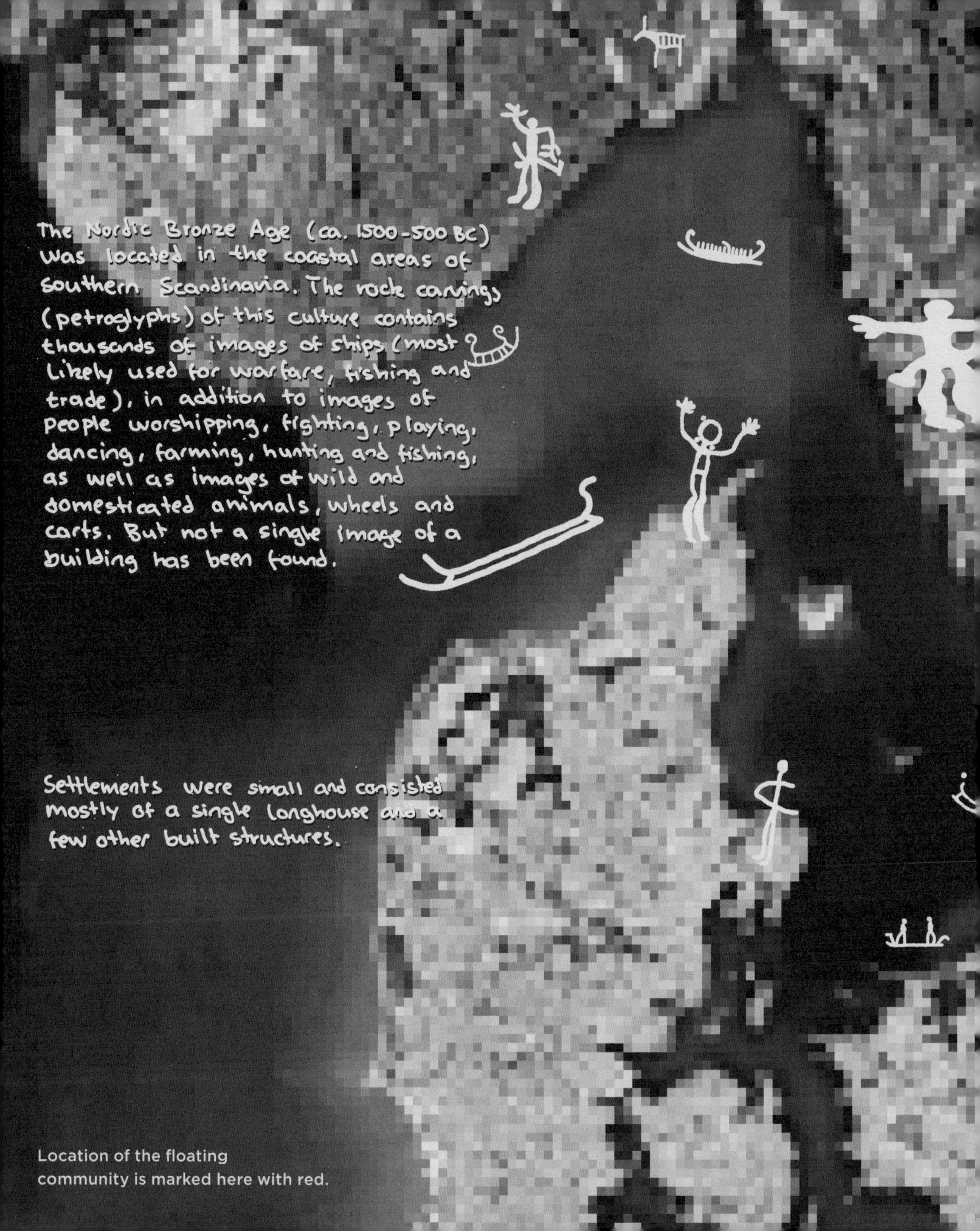

The Nordic Bronze Age (ca. 1500-500 BC) was located in the coastal areas of southern Scandinavia. The rock carvings (petroglyphs) of this culture contains thousands of images of ships (most Likely used for warfare, fishing and trade), in addition to images of people worshipping, fighting, playing, dancing, farming, hunting and fishing, as well as images of wild and domesticated animals, wheels and carts. But not a single image of a building has been found.

Settlements were small and consisted mostly of a single longhouse and a few other built structures.

Location of the floating community is marked here with red.

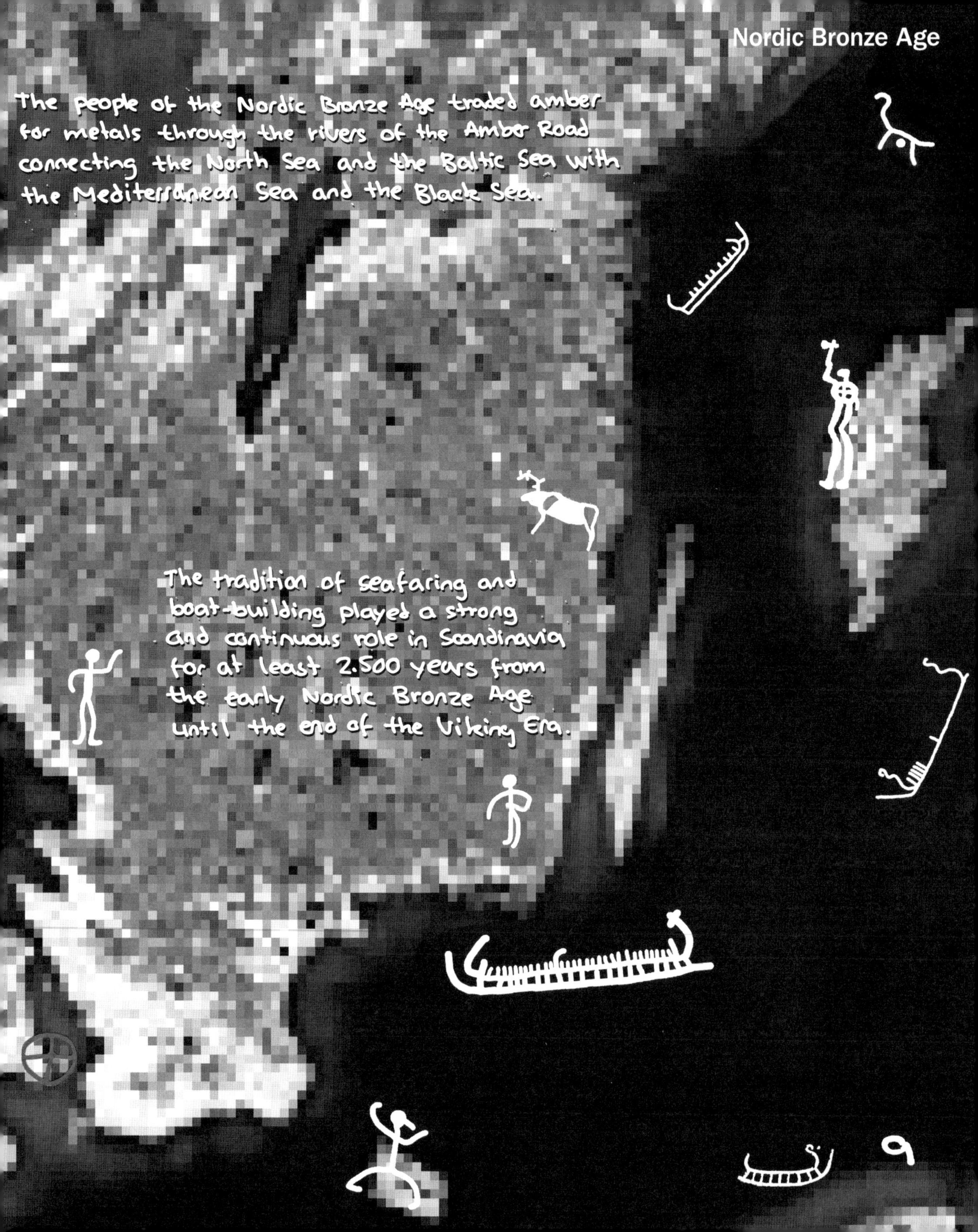

The people of the Nordic Bronze Age traded amber for metals through the rivers of the Amber Road connecting the North Sea and the Baltic Sea with the Mediterranean Sea and the Black Sea.

The tradition of seafaring and boat-building played a strong and continuous role in Scandinavia for at least 2.500 years from the early Nordic Bronze Age until the end of the Viking Era.

"I WANTED TO TRAVEL. HAVING FEW EXPENSES AND EARNING MONEY FROM PERFORMING GAVE ME THE INDEPENDENCE I NEEDED. I HAD A STORY TO TELL AND THERE WAS A WORLD OF PEOPLE LOOKING AT ME. SOMETIMES, I STARTED PERFORMING WITH 600 PERSONS, I TOOK A BREAK AND WHEN I WAS BACK THERE WERE 1,200 STARING AT ME. MY DEFINITION OF SUCCESS HAD TO DO WITH TAKING UP MY BAG AND MY PASSPORT AND PERFORMING WHEREVER I WANTED."

1983, Centre George Pompidou, Paris.

Stephen James Back

"WHAT A CLOWN DOES IS TO PROVOKE THE TRUTH."

11

Since he was 22 years old,
Stephen has been a freelance artist, a clown and a performer.

"Better to have a home,
even if it's little –
everyone should call somewhere 'home.'
Your heart will be wounded
if you have to beg for every meal
from somebody else."

Old Norse poem in Jackson Crawford (ed. and transl.) (2019)
The Wanderer's Hávámal. Indianapolis, Hackett Publishing Company, p. 19.

Postcards
1 = 10 DKK
2 = 50 DKK
3 = 100 DKK

Pirathavnen

[The Pirate Harbor]

15

Tibirke, October 31, 2021

This book exists because Andreas, Gabriel, Kami, Kasper, Örn, Patryk, Sira, Stephen and Storken generously gave me their time and shared their stories, thoughts and experiences with me, and also with Francisca and Frédéric, who helped me. I am deeply grateful to them all, as well as to Per, who initially introduced me to the floating community; its founder Esben, who was my best friend there for a long time; and the former residents, Marie, Lisa and Mette, who helped me to better understand it.

I'm also grateful for the thoughts and perspectives provided by Di (activist), Donald (consultant), Gorm (historian), Jan (architect), John (sociologist), Jonas (lawyer), Lars (documentary director), Michael (engineer), Rasmus (lawyer), Tommy (lawyer) and Uffe (biologist). But above all, I am grateful to Storken, who decided to step in and back me up at a critical point in time when things looked rather difficult.

I cannot say I know Storken very well. Maybe no one does. But I do feel a strong bond and I hope I will come to his rescue, too, should he ever need it.

Although our lives are different, we are both wanderers in search of our own way in life. As

Storken said to me the last time I spoke to him: “Now I’m going out into *my* world”.

I was two years old when my parents found me in a small creek half a kilometer from home. They were in a state of panic, apparently leaving one of my boots behind in the mud, while I was minding my own business.

Storken was not much older when he had to learn how to get to his grandmother three kilometers away, because his parents would often leave him at home without food while they would go out drinking and would forget all about him.

But Storken is a fighter: he is fighting for his right to be here, even though he’s also fighting with his own demons.

I dedicate this book to Storken and to every other member of the floating community, that is to say, the people who do not fit into the modern world’s fixed and ordered way of life.

Henrik Valeur

P.S. While the final layout for this book was being prepared, my friend Örn passed away. He had come to the floating community many years ago, after having lost, first his wife, and then his daughter, and finally his stepson. He maintained an apartment on land but would often fall into a dark hole when he stayed there, so he preferred to live on his dinghies and rafts, where he could also use his skills as a carpenter. He was found dead in his apartment, where he had been lying unnoticed for several weeks. May his tormented soul rest in peace.

IT IS WHAT IT BECO

clusters, while a few are anchored alo
called Fiddler's Green, a floating doc
called the Arrow, and a makeshift pie
used to be a cluster around the rema
this cluster was removed by the auth
community frequently changes as peo
it. The floating community is what it
on individual initiatives that may be
lasting dreams, external pressures or
a collective creativity which is in

MES Most of the floating dwellings are organized in communal

e. Clusters have formed around a raft of old timber called the Cross, a jetty called the Floating Quarter. There also s of a bridge called Donkey Island, but ties in 2020. The configuration of the e move in and out of it – and around ecomes, and this becoming depends iggered by sudden impulses or long-ternal conflicts. These initiatives form ed what gives form to its becoming.

Nordic Wasabi
Plantag
20
mirabelle plum
Rucola
nettle
hunde Rose
soapwort
Lamium album

Mugwort
Mint
Elderflower
cherry
Ribs
dande lion
jelly
Beet root
onion
cress
Apple

In the composting, toilet feces and urine are separated when discharged. Feces are covered with a handful of carbon-rich material, such as sawdust or dried leaves, which is kept next to the toilet, to start the decomposition process and to neutralize the odor.

Composting toilet

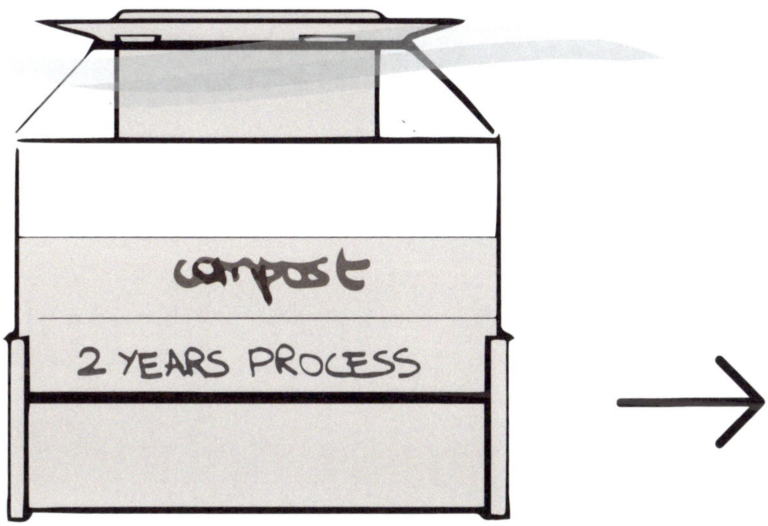

The two buckets containing feces and urine are emptied maybe once a month. Feces are composted for another two years to kill any pathogens before being used to enrich the soil. Urine is stored for a few weeks to kill pathogens and diluted with water before being used as fertilizer.

The composting toilet uses no water and thus saves, on average, 40 liters of water per person per day while avoiding the discharge of polluted water into the environment.

planned and managed societies the wo
has no formal and hardly any *in*formal
may think that they are). And, in contra
open to everyone (because nobody ow
isolated while others spend a lot of t
all the time and sometimes meetings
both visually and audibly in close co
are exposed to the natural forces of
and camaraderie. They share things a
obligations and learn to take responsib
other – not in theory or from a distan

n contrast to the surrounding society, which is one of the most comprehensively has ever seen, the floating community les – and no rulers (even though some to most other communities, this one is the water). Some people live relatively e together. Random meetings happen e arranged. Living on water and being act with each other, the boat people gravitation and evaporation, animosity matters, joys and sorrows. They have ty for their own lives and to help each , but altogether practically, and up close.

"YELL - OOOOHHH!"

stallations at Roskilde Festival

Roskilde Festival

renewable material)

d doors found at a recycling yard

onsisting of wasted wood chips)

27

od stalls and stage at Roskilde Festival

urced from a truck-washing station

Possible locations of – and connections between – floating communities in Denmark with 6m sea-level rise.

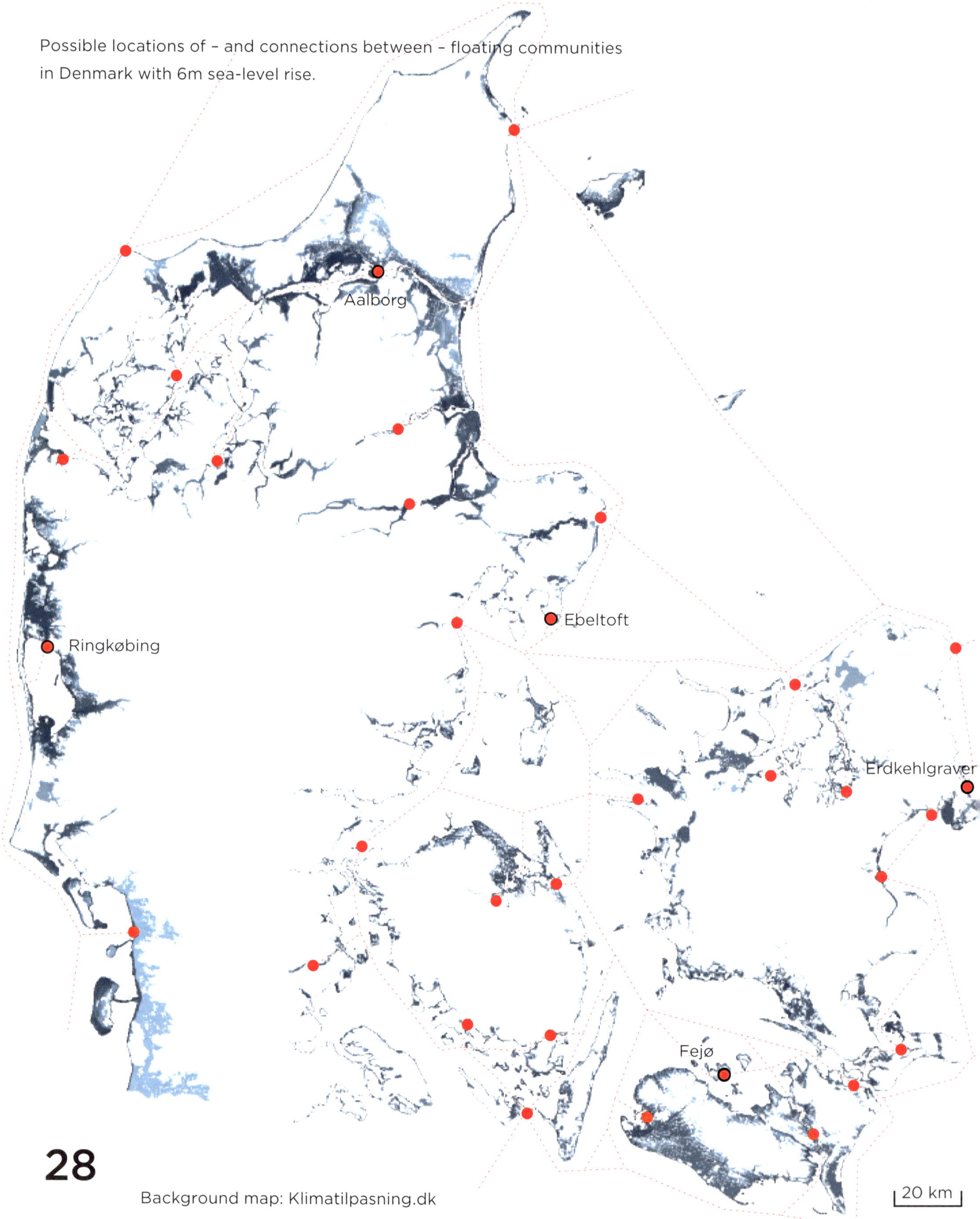

Background map: Klimatilpasning.dk

If floating communities were established in other parts of the country, in the larger port cities and in the more peripheral areas, with people being able to move around between them, new connections could be created across the country. But people would also be able to move around within the individual community and the individual home would typically consist of several smaller units that could also be moved around: the teenager could move away from home, maybe just over to the girlfriend, in his or her own floating room.

Denmark has an unusual lengthy coastline (7.314 km) and a very large number of harbors, relative to the size of the country. Many of these harbors have lost their original purposes, which were related to fishing, trading and ship-building. But many of the facilities related to these activities are still there. Those facilities could be made useful again by floating communities, who need facilities to build and to restore their floating dwellings, to create permaculture on land and sea, and to meet and interact with each other and the local population.

Floating communities may provide accommodation for people who are currently without a home or running the risk of becoming homeless, and people who are already practicing a frugal lifestyle or would like to try doing so.

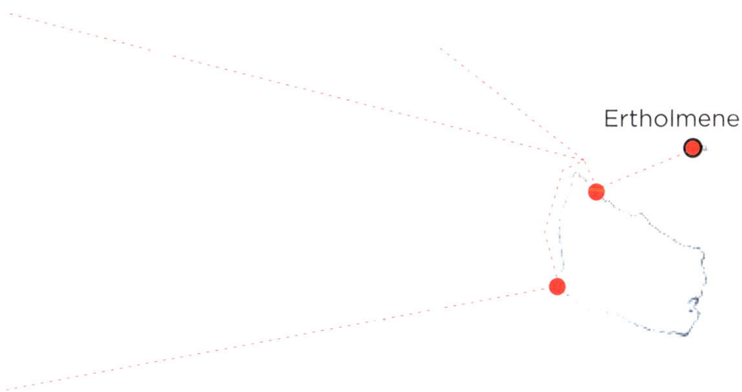

The inhabitants of those communities ought to be able to pursue a nomadic lifestyle, while simultaneously being integrated in the local ecologies they inhabit.

The development of this new form of housing will have to be based on improvisation and ongoing learning, rather than on plans and regulations.

"To think about what it would take to live in a world in which everyone really did have the power to decide for themselves, individually and collectively, what sort of communities they wished to belong to and what sort of identities they wanted to take on – that's really difficult.
To bring about such a world would be almost unimaginably difficult.
It would require changing almost everything.
It also would meet with stubborn, and ultimately violent, opposition from those who benefit the most from existing arrangements."

David Graeber (2004)
Fragments of an Anarchist Anthropology,
Chicago, Prickly Paradigm Press.

"WE ARE NOT HOMELESS, WE ARE HOME-NERVOUS".

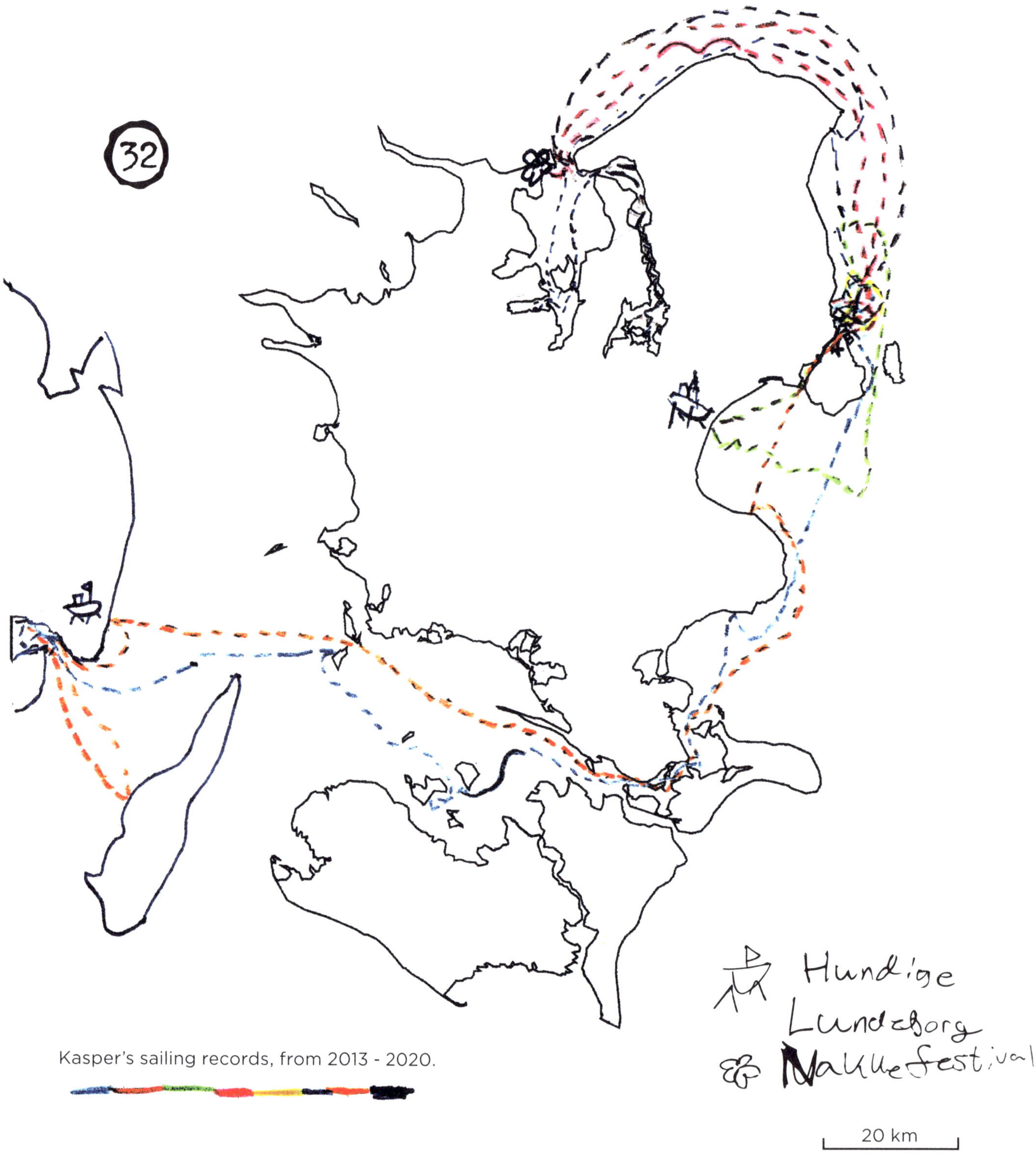

Kasper's sailing records, from 2013 - 2020.

Kasper Hedegreen

The Yaghan people traveled by canoes between the various islands of Tierra del Fuego (named after their many fireplaces). The canoes were made of bark and wood and sectioned with spaces for hunting in the front, paddling in the back and a fireplace midship. All of their tools and utensils, skins and other belongings were stored in the canoe. Waste was thrown into the fire and not into the water, as this was considered to be disrespectful of the sea, which was the source of food and of life itself.

Despite the cold climate, the Yaghans wore little or no clothing but were kept warm by covering themselves in animal grease and huddling around the fireplace. They had also developed high metabolisms that enabled them to generate more internal body heat.

The women would dive into the freezing waters for shellfish, while the men would hunt sea mammals and birds. They used bows and arrows for hunting, with bowstrings made of sea lion tendon, harpoons made of whale bone, darts, daggers and spears, slingshots and snares.

They used bark masks for initiation rites and body paint when visiting other groups — with different colors representing different states of mind.

They erected temporary dwellings close to the sea with a fireplace in the middle and a communal bed of layered branches and dried seaweed, with animal skins on top.

They had a profound sense of individual freedom without any tribal leaders, although the wise elderly had considerable moral influence. They lived in extended families that each had its own territory and were named after that. They roamed around and rarely camped in one place for more than a few days.

The Yaghans died from endemic infectious diseases introduced by the Europeans, to which they had no immunity. Their food sources were depleted by European whalers and sealers, and European ranchers hunted them down.

BEING OUTSIDE,

outsiders, even though they reside in t city of the modern welfare state of Den over a hundred years by the Social De largely been shaped by the same politic which declares itself to be both social a including all of its citizens, there are peop never really invited in. However, from a that outsiders play a vital role in the s members of one species are evolving in

out to be a dead end or to head over

NSIDE The inhabitants of the floating community are

very heart of Copenhagen, the capital

ark – a city that has been ruled for well

ocratic Party, and a state that has very

party. Because even within this society,

democratic, with the stated intention of

who have opted to stay outside – or were

evolutionary perspective, it would seem

cies' chances of survival, because if all

e same direction and that direction turns

cliff, that species might become extinct.

" The first person who, having enclosed a plot of ground, thought of saying this is mine and found people simple enough to believe him was the true founder of civil society.

What crimes, wars, murders, what miseries and horrors, would the human race have been spared by someone who, pulling up the stakes or filling in the ditch, had cried out to his fellow humans:

'Beware of listening to this imposter. You are lost if you forget that the fruits are everyone's and the earth is no one's!' "

Jean-Jacques Rousseau (1755) *Discourse on the origin, and the foundations of inequality among men.* In John T. Scott (ed. and transl.) (2012) *The major political writings of Jean Jacques Rousseau – The two discourses and the social contract.* Chicago, University of Chicago Press, p. 91.

The Utzon Center, an exhibition venue named after the architect Jørn Utzon, is located on the waterfront of Aalborg Harbor. Utzon's father was a renowned boat builder from Aalborg and Jørn Utzon's close association with sailing and the sea is evident in several of his building designs. What would be more fitting than creating a floating community in front of the Utzon Center – as a lived-in experiment with water-based architecture?

Such a community might be established and maintained in collaboration with the local Mathis Shipyard and, apart from being a place for about ten people to reside, some of whom might be architecture students from the nearby CREATE Aalborg University, it could very well include both a floating performance platform on water as well as an outdoor exhibition space on land. Aalborg is located in a fjord named Limfjorden, which is known for its many 'fjord gardens' [in Danish: *fjordhaver*] for shellfish farming, due to an optimal level of salt in the water. The floating community may include a 'fjord garden' on water and a community kitchen on land, where passers-by could have a chance to taste the harvest from the 'garden'.

Aalborg
Perfomance platform
Exhibition
Bicycle parking
Sea farming
Guest shelter
Seating area
Community kitchen
Kitchen garden
Kitchen garden
Composting bins
Storage and workshop
10 m

42

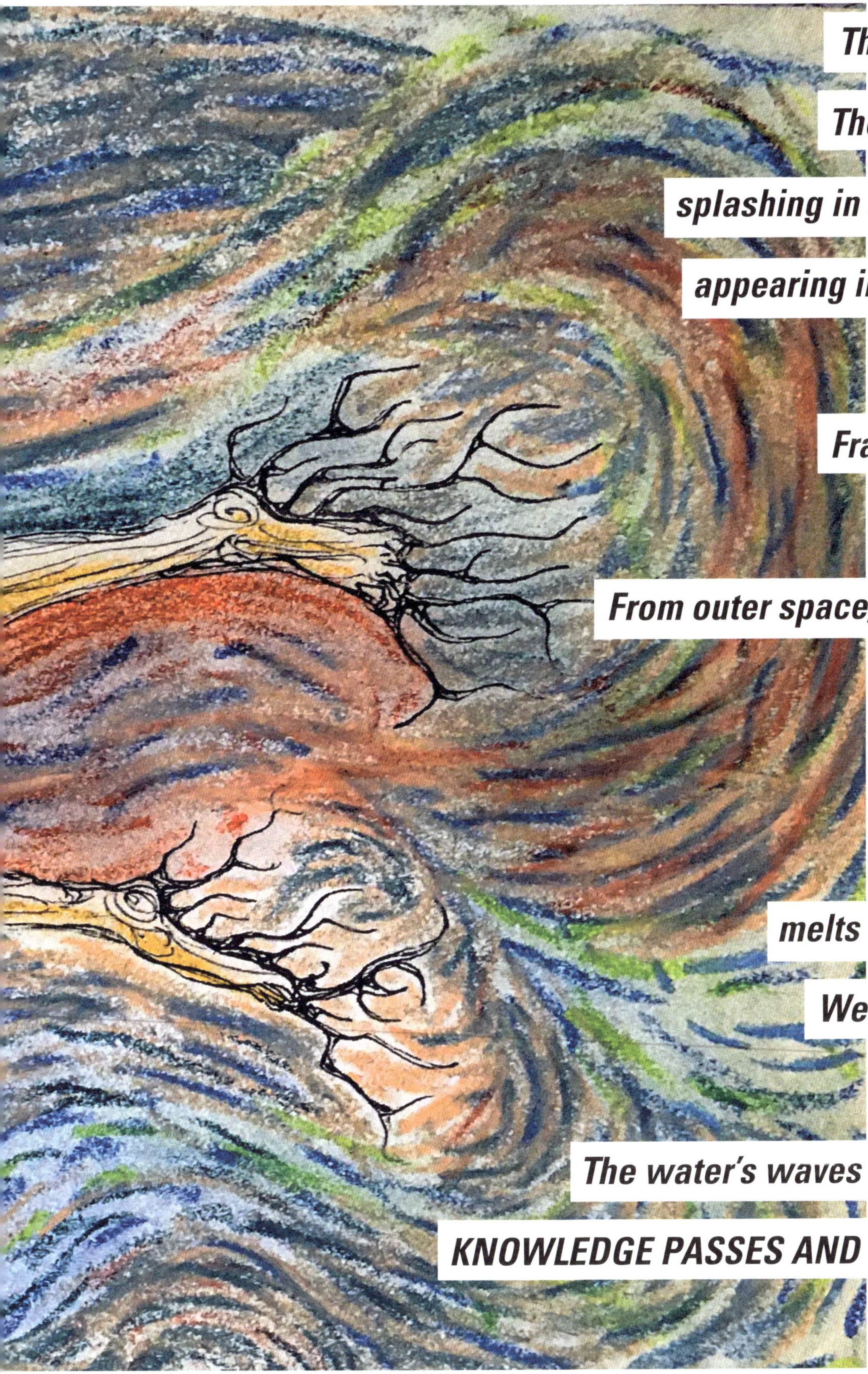

The rhythm of life
The tactile energy
splashing in swirling waves
appearing in the almanac's
confinement
Fractiles invading
the beings
From outer space, in inner space
Maneuvering
eagerly across
the ice, which
melts you understand
We demand action
And ACTION
The water's waves are whirling on
KNOWLEDGE PASSES AND ONLY WISDOM
PERSISTS

ON THE BORDER T
b

and civilized, between fantasy and rea
diagnosed as *borderline*, which is, of co
is here on the border that they are able
inner peace. This is where artists and m
became commodities and the artist a
The border is where the unexpected ar
one becomes the other. What the boa
be showcased in fancy galleries or ent
academic institutions: it is art and magi

 deep and profound sense that mar

boat people inhabit the *borderland* ween land and sea, between wild ty ... Some of them have indeed been rse, one way of marginalizing people. It o express themselves and to find some gicians used to live before art and magic l the magician became a salesperson. the unexplainable may happen, where people create is not art and magic to tainment palaces, or to be analyzed in n a lived, outreaching and interrelating, modern people seem unable to grasp.

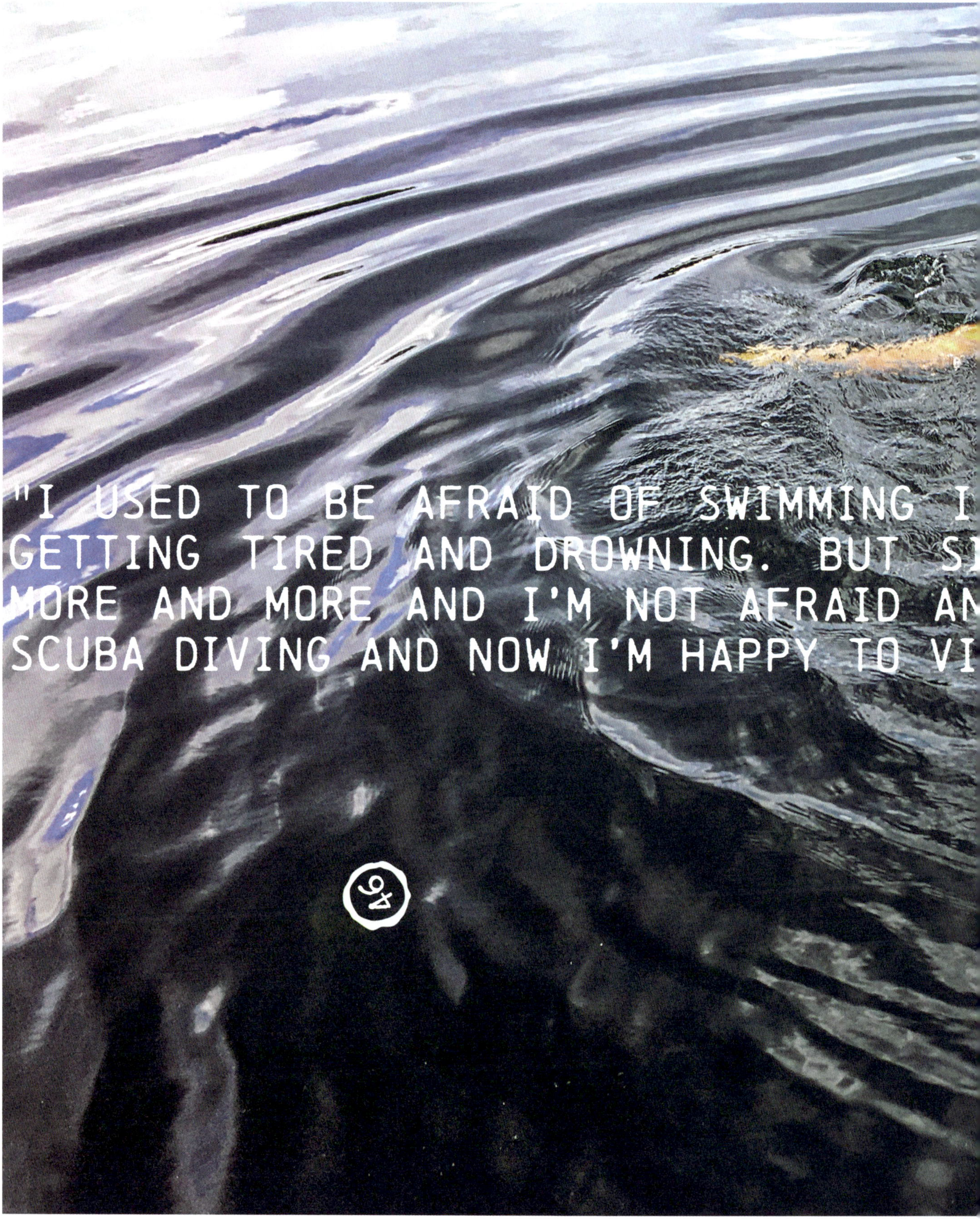
"I USED TO BE AFRAID OF SWIMMING I
GETTING TIRED AND DROWNING. BUT SI
MORE AND MORE AND I'M NOT AFRAID AN
SCUBA DIVING AND NOW I'M HAPPY TO VI

OPEN WATERS BECAUSE OF THE FEAR OF
E I LIVED HERE I STARTED SWIMMING
ORE. I'M REALLY INTERESTED IN DOING
T MY NEIGHBORS BY SWIMMING TO THEM."

A MAGICAL WORL

modern world is an enchanted w
and to accept, as it is (not to be
It is a world in which culture and nat
not separated entities but are clos
and too deep to be explained. Thus,
and the same – wisdom. This is no
where imagination and intelligence
given a psychiatric diagnosis). This i
of modern life. As Leonard Cohen

 There is a crack, a crack in eve

D The alternative to the functional and utilitarian ld, a world to wonder, to respect ansformed into one's own image). re, materiality and spirituality ... are intertwined. This world is too vast nagination and intelligence are one what we typically experience today, re strictly separated (if not, one is he world that appears in the cracks ang: "Forget your perfect offering. thing. That's how the light gets in."

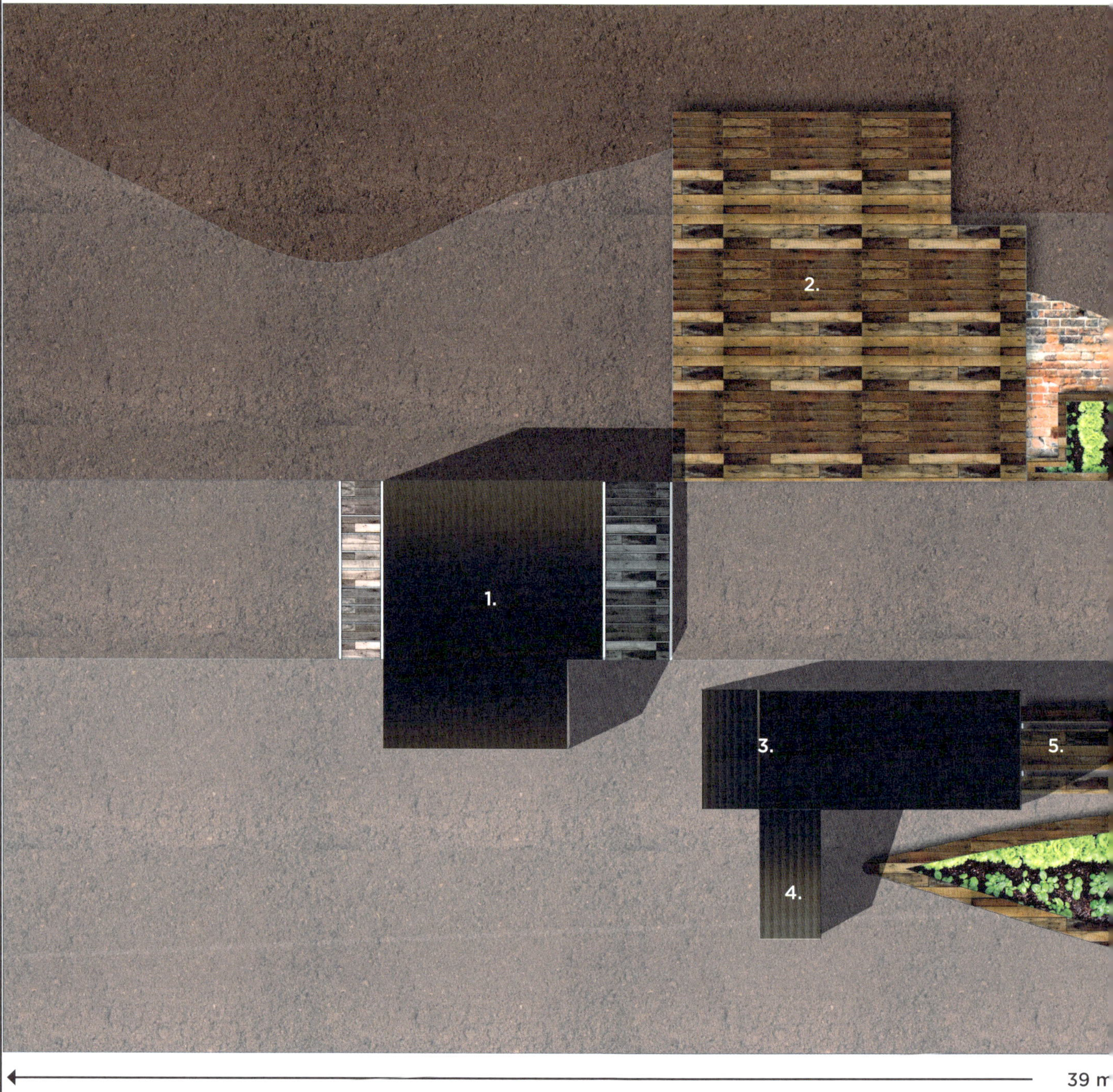

1. Storage space
2. Garden platform
3. Tool shed
4. Composting toilet
5. Benches and table
6. Hen house
7. Sea Wolf
8. Covered seating
9. Fireplace
10. Kitchen garden
11. Composting bins
12. Community kitchen
13. Oven

Community garden

Plan

"WHEN YOU
LIVE ON YOUR
OWN BOAT,
YOU ARE THE
CAPTAIN OF
YOUR OWN
LIFE."

Örn

Örn married and started a family in Iceland. He used to work as a carpenter but the jobs dried up and he began fishing. There were also problems at home, but:

"ONCE I WENT TO SEA I FORGOT ABOUT ALL MY PROBLEMS, I WAS JUST FOCUSED ON MY WORK".

Nevertheless, he decided to move to Denmark in search of a better life. Shortly after, however, his wife died. Then his daughter disappeared and was found dead. And finally his stepson died. A friend told him about some people living in Erdkehlgraven and in an effort to find a new meaning in life he decided to go there on a small boat:

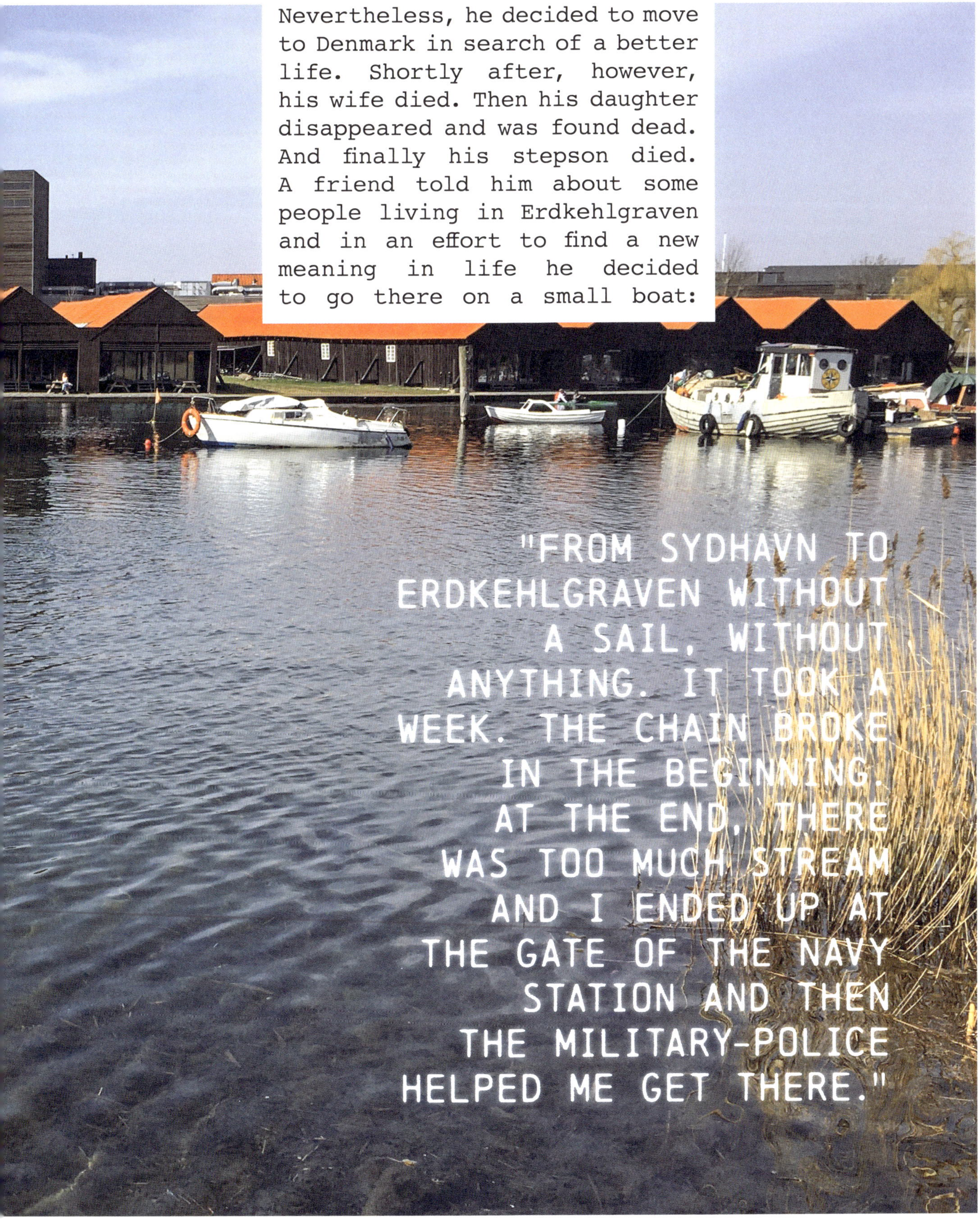

"FROM SYDHAVN TO ERDKEHLGRAVEN WITHOUT A SAIL, WITHOUT ANYTHING. IT TOOK A WEEK. THE CHAIN BROKE IN THE BEGINNING. AT THE END, THERE WAS TOO MUCH STREAM AND I ENDED UP AT THE GATE OF THE NAVY STATION AND THEN THE MILITARY-POLICE HELPED ME GET THERE."

"What is in opposition is in agreement, and the most beautiful harmony comes out of things in conflict".

Heraclitus (6th-5th century BCE),
Fragment VIII. From Aristotle, *Nicomachean Ethics.* In Sweet, D. (1995)
Heraclitus – Translation and Analysis. Lanham, University Press of America.

GRIEF OF LOVE

These days, I was thinking to myself

That I would walk along the quiet streets of Oslo, between railway tracks and graveyard

That I would come to you, down the valley, at the foot of Ekeberg

I would open your door, or first you would look out your window

My longing shows me the way

Perhaps it would snow

Huge white fluffs below the sky above me. Nothing is as quiet as snow

All would be as if it had been forever

The world is quiet. Don't utter a word

Outside is a world, right now, which was

We can create something greater than us, greater than we can grasp, greater than words

Are you breaking the silence now

Will we remain trapped in nemesis

The falling fluffs are hope and they fall down softly and create a blanket

A blanket of hope, I step on:

Quiet night. Be quiet now. It's all right I'm here with you. And I believe we share the same origins.

HISTORIEN : DER VAR ENGANG EN LILLE
ENSOM DRENG. HAN VAR IKKE ENSOM
59

DREAMING

We have talk in other ha involved and made a study trip. We for all the neighbors centered on ho preserve the heritage and protect the a permacultural laboratory and we h plant to join in. We have talked about to the parliament for the purpose of would take at least a month to do th launching a political party for anar fight for the community. So far, none

doesn't matter, because as long

about creating floating communities
ors – we even got several people
ave talked about arranging a festival
to avoid environmental degradation,
vildlife. We have talked about creating
e invited the nearby waste-to-energy
lking from the other end of the country
hering support for the community – it
we reckoned. We have talked about
ists – a contradiction in terms – to
f this has been realized. But it really
s we're able to dream, there is hope.

"A STREET ARTIST FRIEND DREW THIS ON MY TINY F
OLD WOLF, CARESSING A SHEEP. I THOUGHT FOR A LO
A FAMILY ARE THE ONLY RESPONSIBLE THINGS TO DO

.TING HOME. IT DEPICTS ME AS A RADIATING & WORN
TIME THAT STAYING A LONE WOLF AND NOT CREATING
A WORLD OF OVERPOPULATION & OVERCONSUMPTION."

Örn started out with a small boat,
constructing an enclosed space on top of it,
found a raft ...

1.

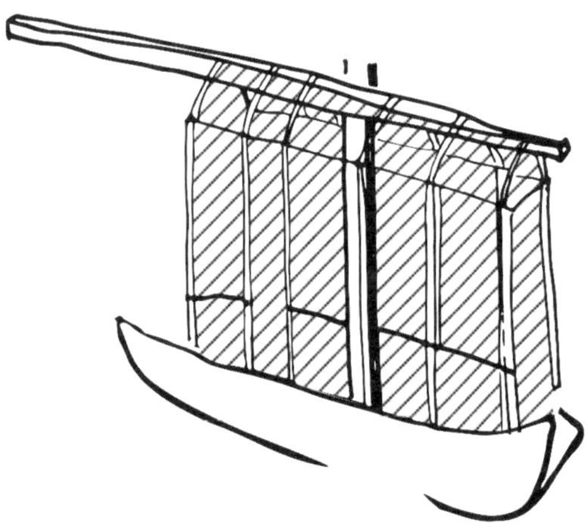

2.

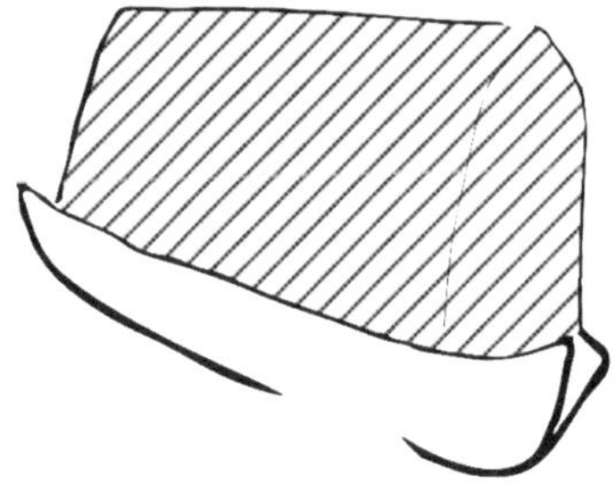

3.

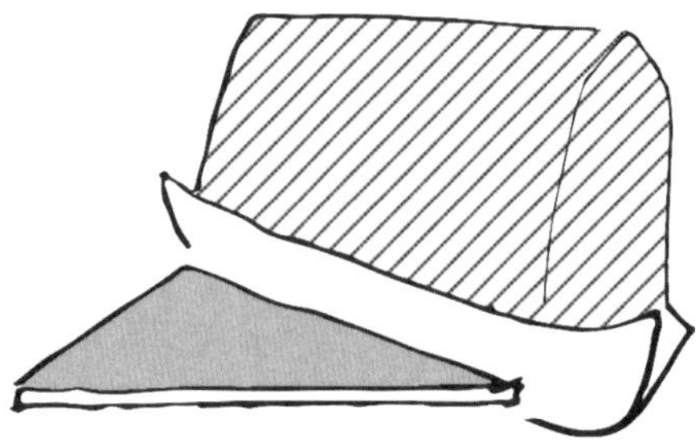

4.

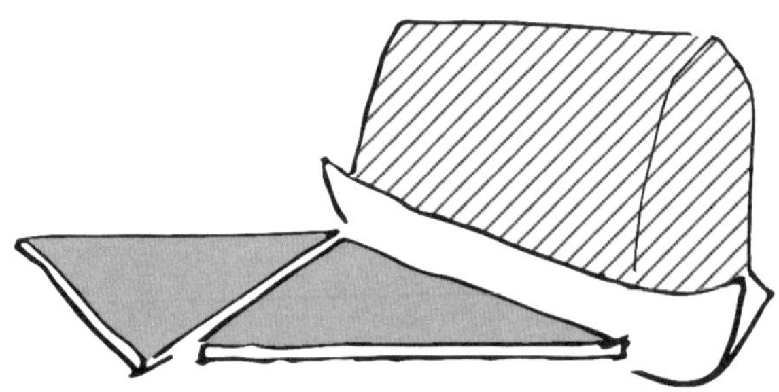

5.

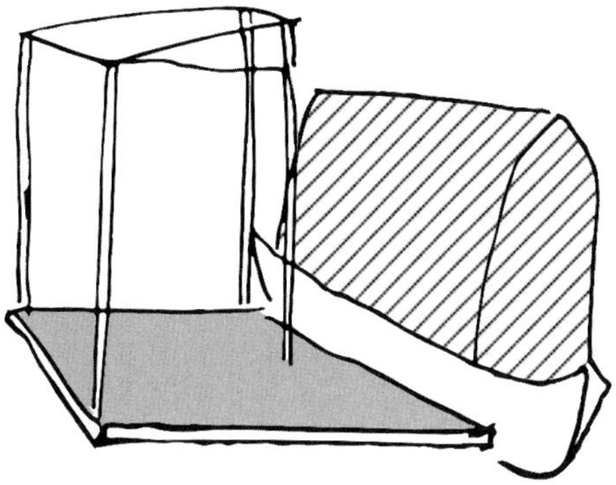

6.

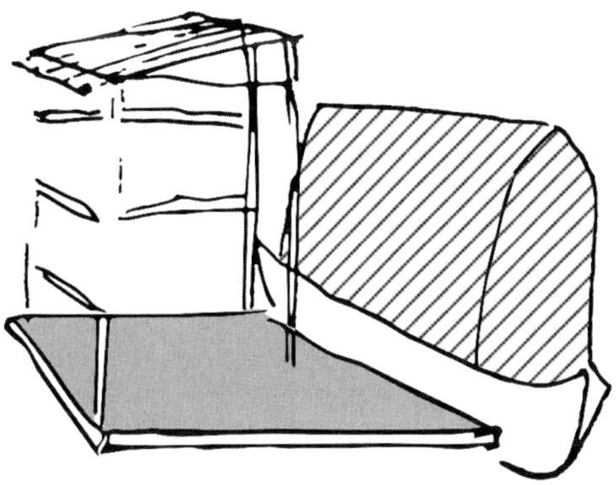

7.

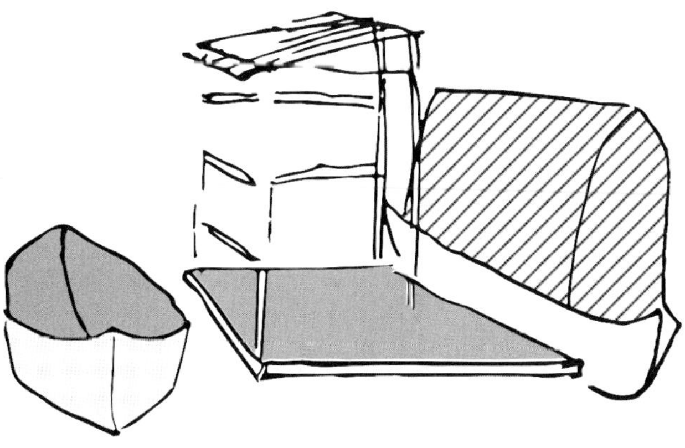

8.

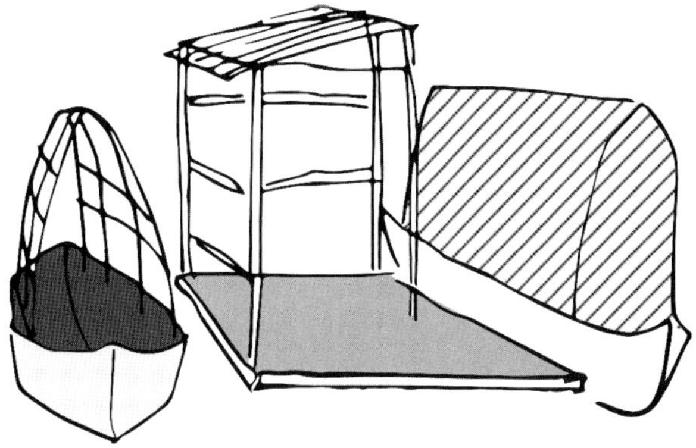

9.

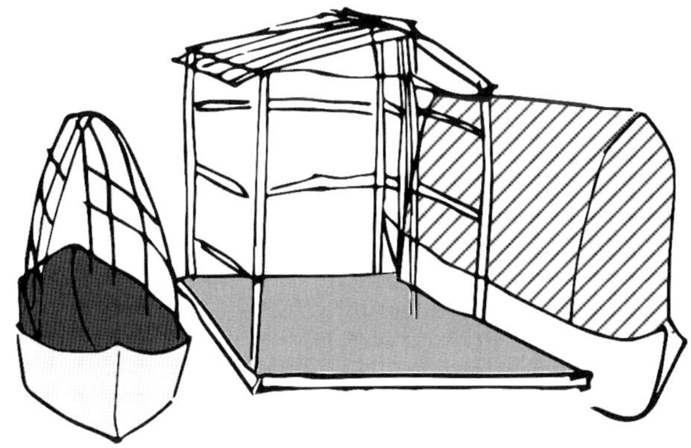

10.

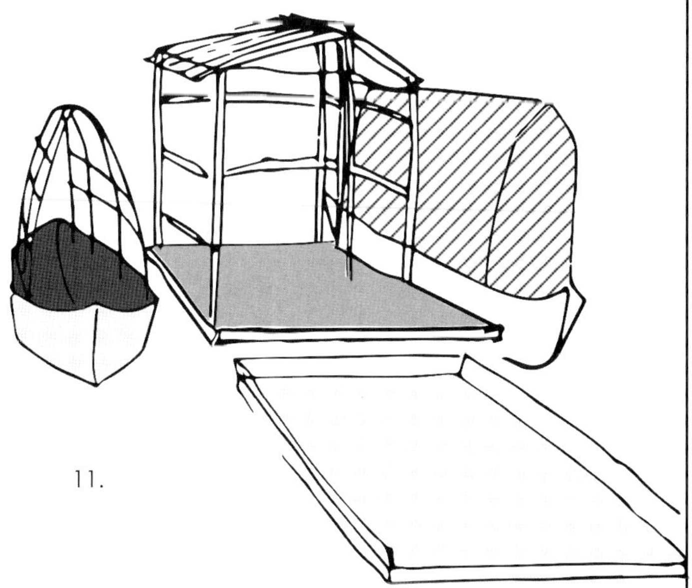

11.

2006 - 2011

2011 - 2012

2012 - 2013

2013 - 2014

2014 - 2016

2016 - 2018

2018 - 2019

50 m

”O flight of swans!
Before me, you have undone the lid of silence tonight.
Beneath this muteness I hear
In the void, in water and on land
A similar fluttering of impetuous and agitated wings.
Vegetations
Are bustling wings below their earthen sky,
Below the earthen obscurity, indifferent to
destination,
Unfurl wings of sprouts,
A flight of thousands of seeds.”

Rabindranath Tagore (1914)
Balaka, “Flight of Swans”, No.36. In *Prithwindra Mukherjee* (ed. and transl.)
Rabindranath Tagore – Selected Poems III

Dybvig Havn is a small old natural harbor located at the end of a navigable channel in the shallow waters on the southeastern side of the island, Fejø, where it is sheltered from the prevailing western winds and the cold winds from the north. It was initially used for fishing, but after construction in 1814, it became a shipping harbor for agricultural produce being exported from the island. Trading facilities, craft businesses and small-scale industry evolved around the harbor, including a shipyard established in 1878 that still exists, although it is only being used today for repair works. After another harbor was established on the island in 1920, Dybvig Havn slowly began to loose its importance.

To the east of the harbor is a groyne of stones on which a pier could be constructed, which would create a suitable location for a floating community. On land there is a plot of terrain where various common facilities for the community could be established. There is a grocery store within walking distance and the existing shipyard, with its slipway, could be used for maintenance work on the floating dwellings.

Fejø
Bicycle parking
Kitchen garden
Storage and workshop
Community kitchen
Exhibition
Seating area
Composting bins
Theater
Guest shelter
10 m

With materials found either at the architecture school or in dumpsters around the city, Örn built a composting toilet and a rainwater-based sink.

BOAT
MOVING
PARTY

"PLEASE DON'T TAKE MY HOUSE AWAY FROM ME".

FLOATING

The floating community
are temporary and re
literal sense of being located on water, but
exact size or plot of land, or by a fixed form
shifting hands, layout and position. The floa
its future being uncertain and insecure, due
I have a home tomorrow?"), and in the sen
life escapes conventional definitions and c
than being grounded on land, is an easy w
global warming), the scope and speed of wh
maintain social distancing, during a pande
other and with nature. Moreover, changes in
accommodated when dwelling parts can b

parts: the teenager who is moving out ca

anges the place it occupies, but the changes sible. As a settlement, it is floating in the so in the sense of not being defined by an content. Individual dwellings are frequently g community is also floating in the sense of the pressure exerted by the authorities ("will of being inhabited by people whose way of ined conventions. Floating on water, rather of adapting to rising sea levels (caused by n remains unknown. It also makes it easy to c, while staying in close contact with each e individual household set-up can easily be dded or subtracted without affecting other simply steer his bedroom to somewhere else.

The large, open umiak was rowed by women and used for transport and to hunt whales and walrus.
The little, light qajaq was used by men to hunt seals and other animals at sea.
A tupilaq was made in secret by the shaman and placed in the sea in order to seek out and destroy an enemy. But if the enemy had greater magical powers, the tupilaq could be turned against its maker.
During spring, the roof of the peat hut would be dismantled and the driftwood and the skin would be carried inland.
During summer, individual families would live inland, in tents made of driftwood and skin.
During summer, people would hunt caribou and gather berries inland

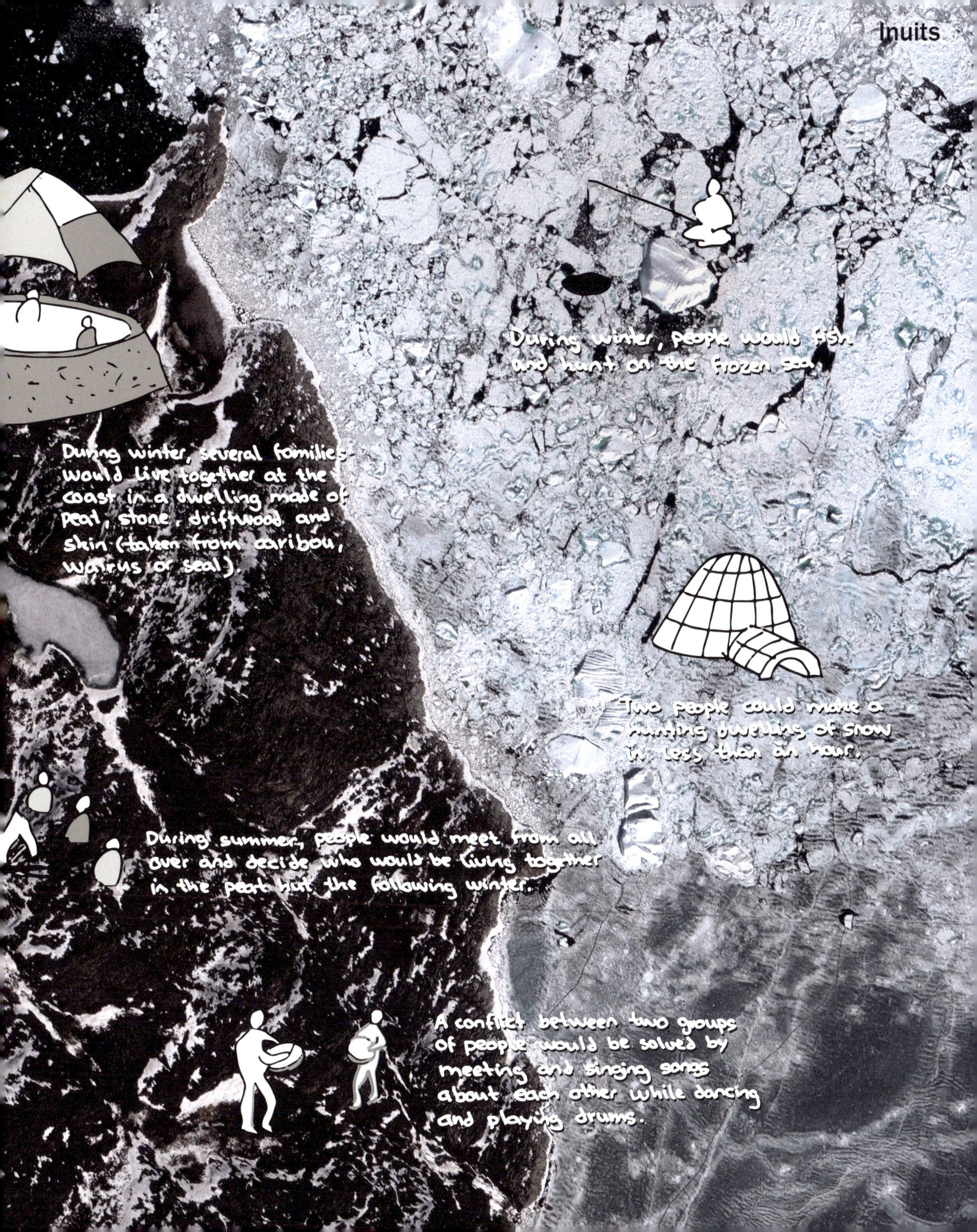
During winter, people would fish and hunt on the frozen sea.
During winter, several families would live together at the coast in a dwelling made of peat, stone, driftwood and skin (taken from caribou, walrus or seal).
Two people could make a hunting dwelling of snow in less than an hour.
During summer, people would meet from all over and decide who would be living together in the peat hut the following winter.
A conflict between two groups of people would be solved by meeting and singing songs about each other while dancing and playing drums.

Ash-based laundry detergent

Potassium-rich ash, from wood, ferns, banana skins, potato peels, etc., is filtered through a sieve and mixed with (rain-)water in the ratio of 10 grams to 1 deciliter for one batch of washing. The mixture is left for 2 days and stirred occasionally before being filtered through a fine cloth. The remaining liquid should be clear and yellow, and without sediments. Essential oils can be added for fragrance. The leftover ash from the filtering can be added to the compost.

This detergent causes no pollution to the environment.

"I CLEAN MY CLOTHES BY HANGING THEM OUTSIDE FOR A FEW DAYS OF RAIN AND WIND."

"WHATEVER FOOD I DO NOT CONSUME, I LEAVE OUTSIDE FOR THE BIRDS TO TAKE. WHATEVER IS LEFT, I PUT DOWN INTO THE COMMON COMPOST."

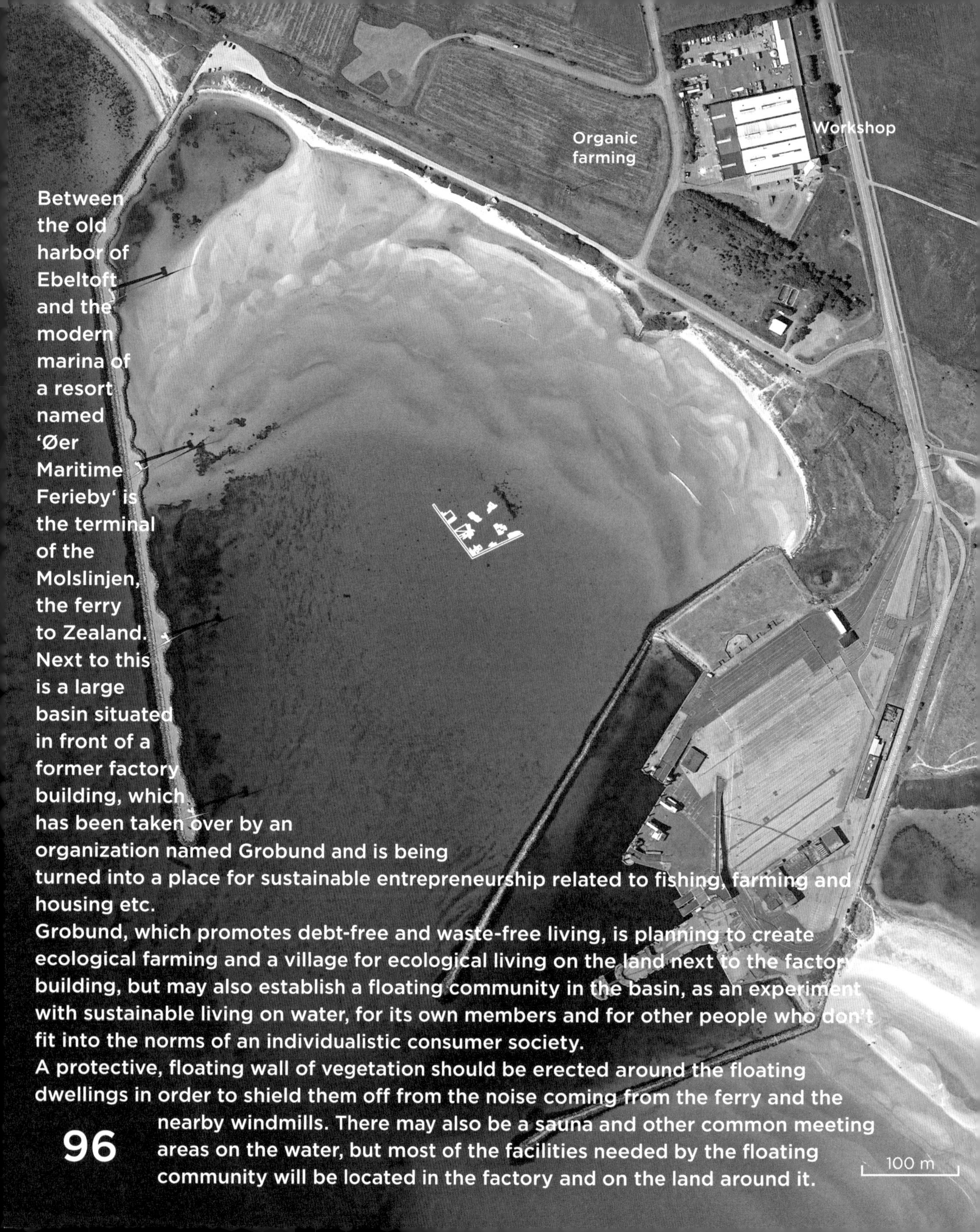

Between the old harbor of Ebeltoft and the modern marina of a resort named 'Øer Maritime Ferieby' is the terminal of the Molslinjen, the ferry to Zealand. Next to this is a large basin situated in front of a former factory building, which has been taken over by an organization named Grobund and is being turned into a place for sustainable entrepreneurship related to fishing, farming and housing etc.

Grobund, which promotes debt-free and waste-free living, is planning to create ecological farming and a village for ecological living on the land next to the factory building, but may also establish a floating community in the basin, as an experiment with sustainable living on water, for its own members and for other people who don't fit into the norms of an individualistic consumer society.

A protective, floating wall of vegetation should be erected around the floating dwellings in order to shield them off from the noise coming from the ferry and the nearby windmills. There may also be a sauna and other common meeting areas on the water, but most of the facilities needed by the floating community will be located in the factory and on the land around it.

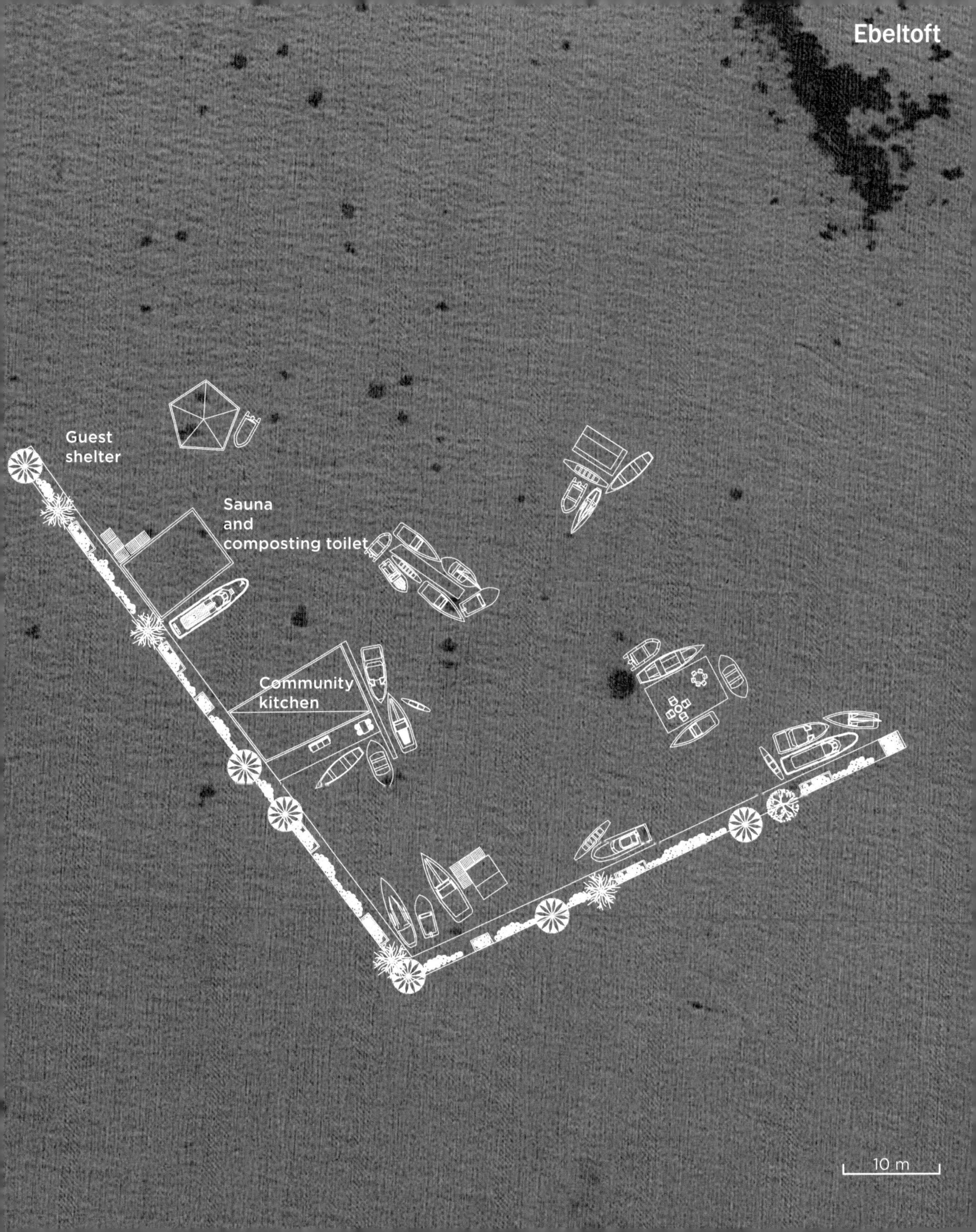
Ebeltoft
Guest shelter
Sauna and composting toilet
Community kitchen
10 m

"Freedom is not an ideal located outside of man; nor is it an idea which becomes myth. It is rather the indispensable condition for the quest for human completion."

Paulo Freire (2005 [1970])
Pedagogy of the Oppressed. New York, Continuum, p. 47.

Ethnic Vietnamese people have lived in Cambodia for generations, but most of them are stateless and therefore not allowed to own land. Moreover, they have often been victims of xenophobia and persecution and have traditionally opted to live in villages of boathouses or houses on stilts in lakes and rivers.

The floating village of Kompong Luong, located in the Tonlé Sap Lake, consists of boathouses for more than a thousand households, as well as floating pagodas, petrol stations, shops and workshops, clinics, restaurants, bars, a church and a police station, as well as chicken coops and fishponds.

The boathouses each have a multipurpose veranda that can be used for cooking, selling things, resting or cutting someone's hair.

Kompong Luong
In the dry season, when the water levels drop, the whole village is towed further into the lake.

BUILDING ON SUNKEN

islands was created across from the Citad
Copenhagen. These islands were made of tra
foundations constituted by sunken ships. The
served as the main base for the Danish na
ramparts with cannon batteries and gunpow
floating community is located in a moat cal
is still possible to find very large pieces of
barrier that the navy constructed in order to
of the fleet was sacked by the British and ir
naval base was closed in 1993, the area b
and many historic buildings have been retr
educational institutions, including the schoo

well as being retrofitted to accommoda

SHIPS Beginning in the late 17th century, a string of artificial
at the northern entrance to the Harbor of
stones and mud that were heaped on top of
rchipelago, which became known as Holmen,
for three centuries. It included moats and
er magazines, barracks and a shipyard. The
d Erdkehlgraven. In the surrounding reeds, it
nber, that were probably parts of a floating
ecure the fleet. Nevertheless, in 1801, much
807, the entire navy was captured. After the
came a terrain for extensive redevelopment,
tted to accommodate a number of creative
of architecture, film and performing arts, as
a number of upscale residential properties.

NN came to Denmark in search of freedom and love. With his knowledge in mechatronics he volunteered at Guldminen, where scrap materials were given new uses. It was:

ALLEMANDSRETTEN
EVERYMAN'S RIGHT
FOODSHARING
COPENHAGEN

Storken used to live in an outside broadcasting van that had once belonged to the Danish Broadcasting Corporation (DR), and which is now parked on the road next to his boat.

Inside the van is full of memorabilia from his life as a street artist.

A COMMON GAR

common garden where herbs a
connection with this, an improm
very same grounds, albeit on th
rampart, is the exclusive restaura
attracts its individual audience, b
principles and contains some of
find on either side of the rampa
Noma's cuisine can be said to
boat people the creativity i

DEN On land, the boat people have created a

d vegetables are grown, and in u kitchen and a campfire. On the other side of a small overgrown , Noma. Each of the two kitchens . each is based on similar culinary e same raw ingredients that they While the creativity unfolding in e subject to abundance, for the rather dependent on necessity.

110

A clay pot is placed inside a bigger clay pot and the space between them is filled with sand, which is kept moist by occasionally adding (sea-) water. The pot-in-pot refrigerator is covered by a lid or a wet cloth and placed in a dry and ventilated space. As the water evaporates, the thermal reaction lowers the temperature of the inner pot, up to 10 degrees Celsius below the outside temperature.

Pot-in-pot refrigerator

The pot-in-pot refrigerator is most effective in hot and dry climates.

"The formula of today's architecture is nothing but the satisfaction of the sedentary and residential needs of the citizen who seeks nothing but peace and quiet and desires a certain level of comfort (central heating, refrigerator, washing machine, TV set, etc). [It is] architorture".

Raoul Hausmann (1967) In Vostell, W. and Higgins, D. (eds.) (1969) *Fantastic Architecture.* New York, Something Else Press.

On one of his boat trips, Kasper learned to caulk his boat with pitch, i
the traditional way, so he's now able to help others caulk their boats

Kasper Hedegreen

OFF-GRID

The boat people
they are not con
heating, sewage system or water sup
breakdowns in these systems or oblige
electricity available is what comes out
solar panel, and while water has to
carried back in containers, careful co
being used – and how they might be re
their own composting toilets. It only ta
and a canister to separate excrement
going and it has to be emptied, perhap
water and excrements are composted.

waste, these elements are eventu

e living off-grid. This means to say that
cted to the public power grid, district
y, and therefore not affected by any
to pay for their maintenance. While the
f a small solar power bank or a single
collected half a kilometer away and
ideration goes into how resources are
sed. The boat people have constructed
es a piece of wood to sit on, a bucket
rom urine. Some sawdust will keep it
once a month. The urine is mixed with
gether with composted organic kitchen
ly used to fertilize the kitchen garden.

"YOU CARE ABOUT WHAT IS THERE,
NOT WHAT YOU WANT TO BE THERE".

Kasper Hedegreen

Drawing by Matilde Bordessoule

Kære Sk
Mennes
Du indviteres hermed til Påske, meditation og yoga. Skærtorsdag og Langfredag. Vi vil give appetiser på vores uddannelse 'Namaste'. Vi viser forskellige teknikker til at arbejde med værdier, omsorg, overbevisninger, humor, kreativitet, relationer, kommunikation, vaner og forandring. Det koster 500 kr pr snude. Tag sovepose/underlag med eller tjek ind på værelse. Husk håndklæde og sengelinned.
I dit tilbageblik på livet vil du se, at de øjeblikke hvor
sandhed har levet, er de øjeblikke, hvor du har handlet i kær
ånd.

121

"I TRY TO DIGEST INFORMATION, ACQUIRE KNOWLEDGE, AND PERCEIVE WISDOM. IN CONSULTATION WITH A CLIENT, MY INTENTION IS TO INSPIRE HIM OR HER WITH MY SELF-CHOSEN TRUTH, THAT WE CAN COMMIT TO THE MIRACLE OF OUR BODIES' IMMUNE SYSTEM, USING BREATHING TECHNIQUES, AN INFORMED DIET, CONSCIOUS MOVEMENT, AND L O V E ."

The Uros live on floating islands on Lake Titicaca, 3.812 meters above sea level. The islands are built from blocks of Totora roots and bundles of Totora reeds and anchored to eucalyptus poles rammed down into the bottom of the lake. New layers of reeds are constantly added for purposes of replacing the layers that rot away, especially during the rainy season. When the islands were first built, several centuries ago, they were meant to provide refuge from the domination of other peoples like the Tiwanaku, the Qullas and the Incas. They used to be located in the middle of the lake, about 15 kilometers from shore, but were recently moved to the shallow waters near the coast.

The houses on the islands and the balsa boats are also made of Totora reeds, which are also used as medicine, used to make tea and used to eat. Apart from that, the Uros are fishing and hunting birds on the lake. They even have domesticated local water birds that can catch fish for them.

Solar panels provide electricity for televisions and other modern appliances. And the islands have become one of the main tourist attractions of Peru.

The International Space Station is operated by multiple nations, in cooperation. It is made up of modules, components and elements, which have been shipped from Earth and put together in space since 1998. Modules can be added to or removed from the existing structure. Spacecrafts, with crews and cargo, regularly dock at the space station.

The space station contains various life support systems, including atmospheric control systems producing oxygen and removing carbon dioxide and methane, solar panels providing electricity, insulation, radio transmission via antennas and satellites, Linux operating systems, water supply systems, food supply facilities, sanitation and hygiene equipment (feces and urine are separated and urine is treated and recycled into drinking water).

It has been demonstrated that urea, the second most common compound in urine after water, can be mixed with moon dirt and used as a building material, a geopolymer, which is similar to concrete and could eventually be used to build habitats on the moon.

Various plants are grown in the space station for the psychological well-being and physical health of the crew, as well as for research purposes.

SUSTAINABILITY

The
deve

its administrative and judiciary appara
is sustainable. The floating community
and inclusive, and because people ar
help each other. And it is environmenta
very little: some people even try to li
little, when compared to the rest of
nature: for instance, by creating safe
Thus, according to an independent bi
Copenhagen's Inner Harbor for many ye
where the waterfowl breed. And accord
waterfowl has increased since th

te claims to be the patron of sustainable pment. But at the same time, through s, it is trying to destroy something that socially sustainable because it is open trying to take care of themselves and sustainable because people consume without money, and thus pollute very ociety, while they also try to protect esting areas for the local waterfowl. ogist, who has studied the bird life in s, this is the only place in Erdkehlgraven to this expert, the number of breeding floating community was established.

Andreas consumes 8 m^2 living space compared to 85 m^2 for the average Dane.

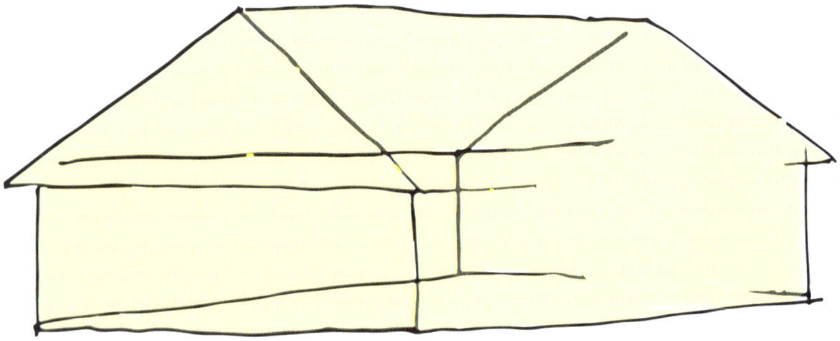

Andreas consumes 20 l of drinking water per week for cooking and washing compared to 769 l for the average Dane.

"MINIMIZING USE IN ORDER TO REDUCE CONSUMPTION AS MUCH AS POSSIBLE"

Andreas uses energy for his phone (6 kWh/year) and his laptop (50 kWh/year), a head torch and a refillable oil lamp for lighting, and a fire stove with scrap wood for heating. No refrigerator (preferring not to hoard food) or any other electrical amenities. The average Dane consumes about 10,000 kWh per year.

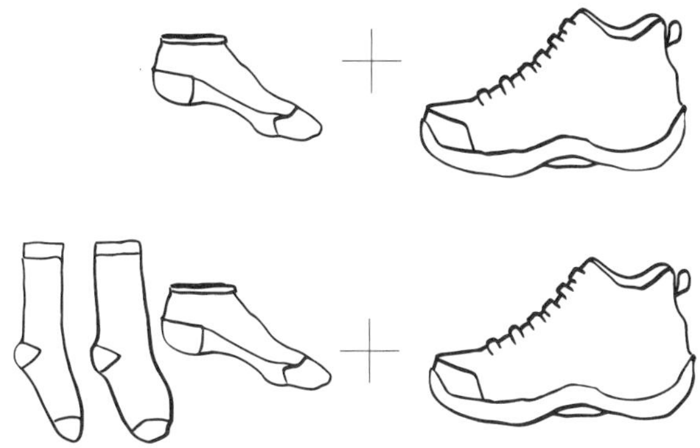

Andreas uses the same shoes summer and winter – with a pair of extra warm socks in winter.

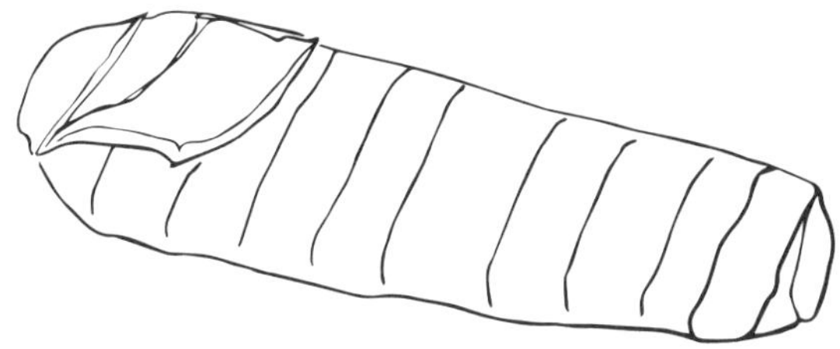

Andreas has used the same sleeping bag for travels and at home for 13 years.

Andreas uses a camping stove both for travels and at home.

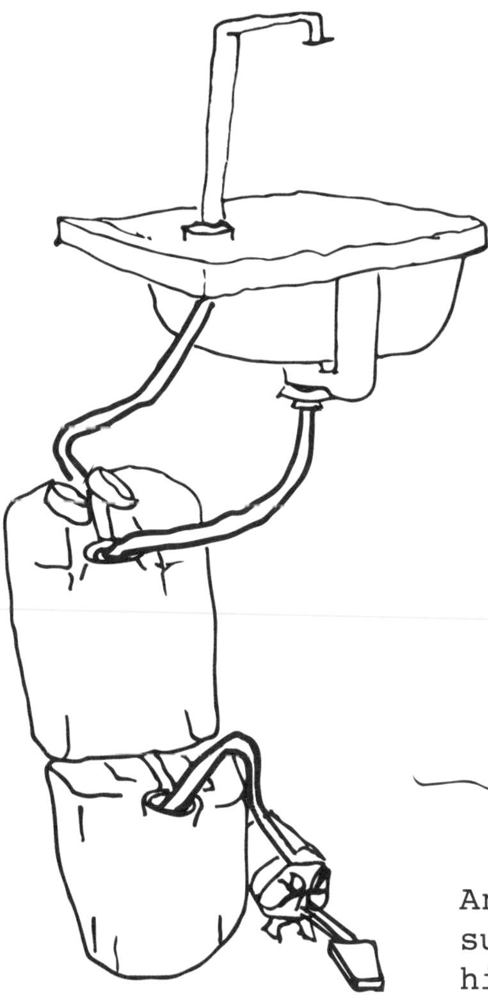

Andreas takes a shower every morning by jumping into the water and using biodegradable soap.

Andreas has designed his own renewable water supply system with a foot pump, which gives him better control over his use of water.

At the end of February 2021, the city of Copenhagen experienced severe col
and the harbor froze for several weeks. After being isolated on his boat fo

everal days, he decided to break through the ice, swinging a rope with two ricks attached, in order to get to shore so that he could fetch drinking water.

"Wabi-Sabi is the quintessential Japanese aesthetic.
It is a beauty of things imperfect,
impermanent, and incomplete.
It is a beauty of things modest and humble.
It is a beauty of things unconventional ...".

Leonard Koren (1994)
Wabi-Sabi for Artists, Designers & Philosophers.
Point Reyes, Imperfect Publishing, blurb.

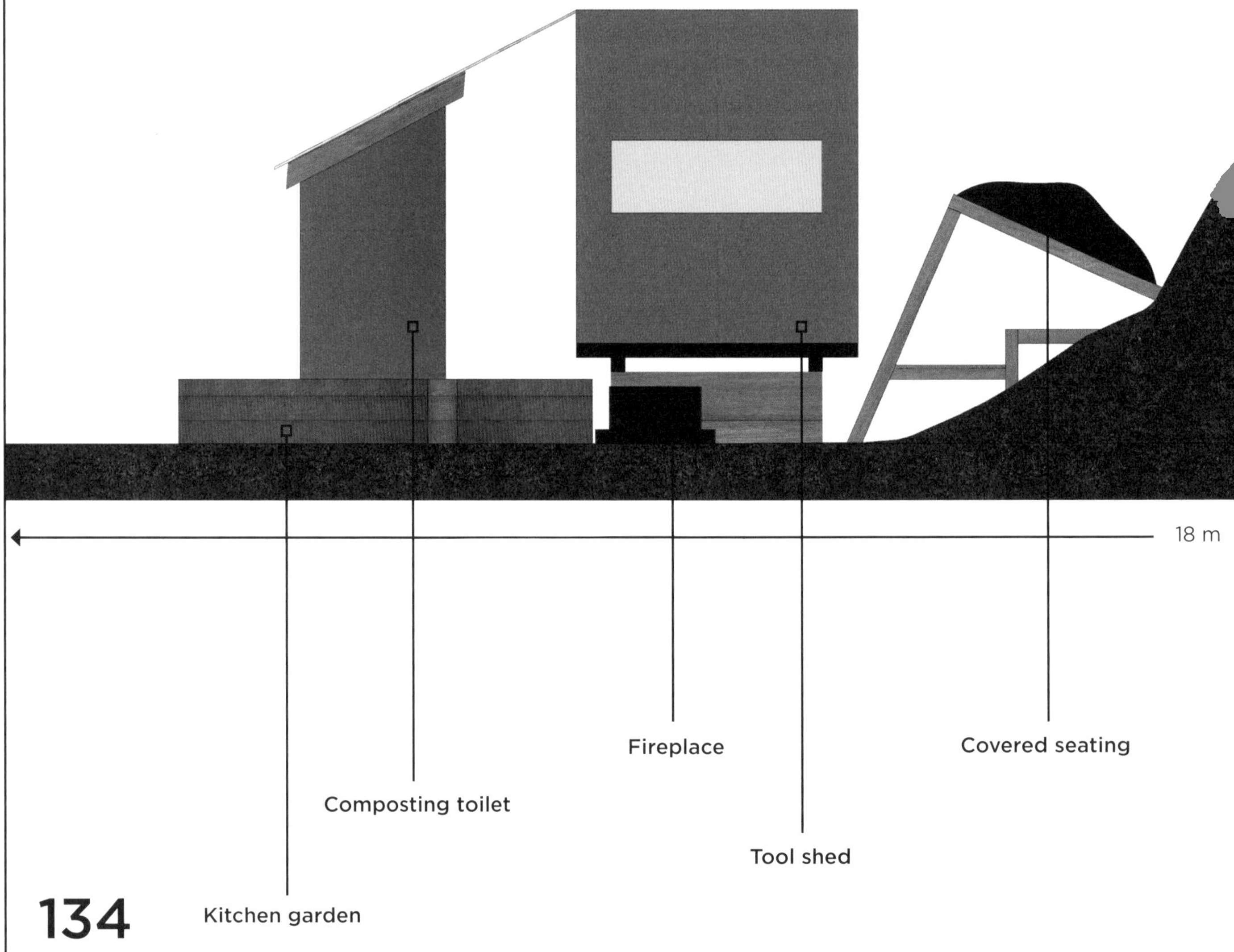
18 m
Fireplace
Covered seating
Composting toilet
Tool shed
Kitchen garden

Community garden

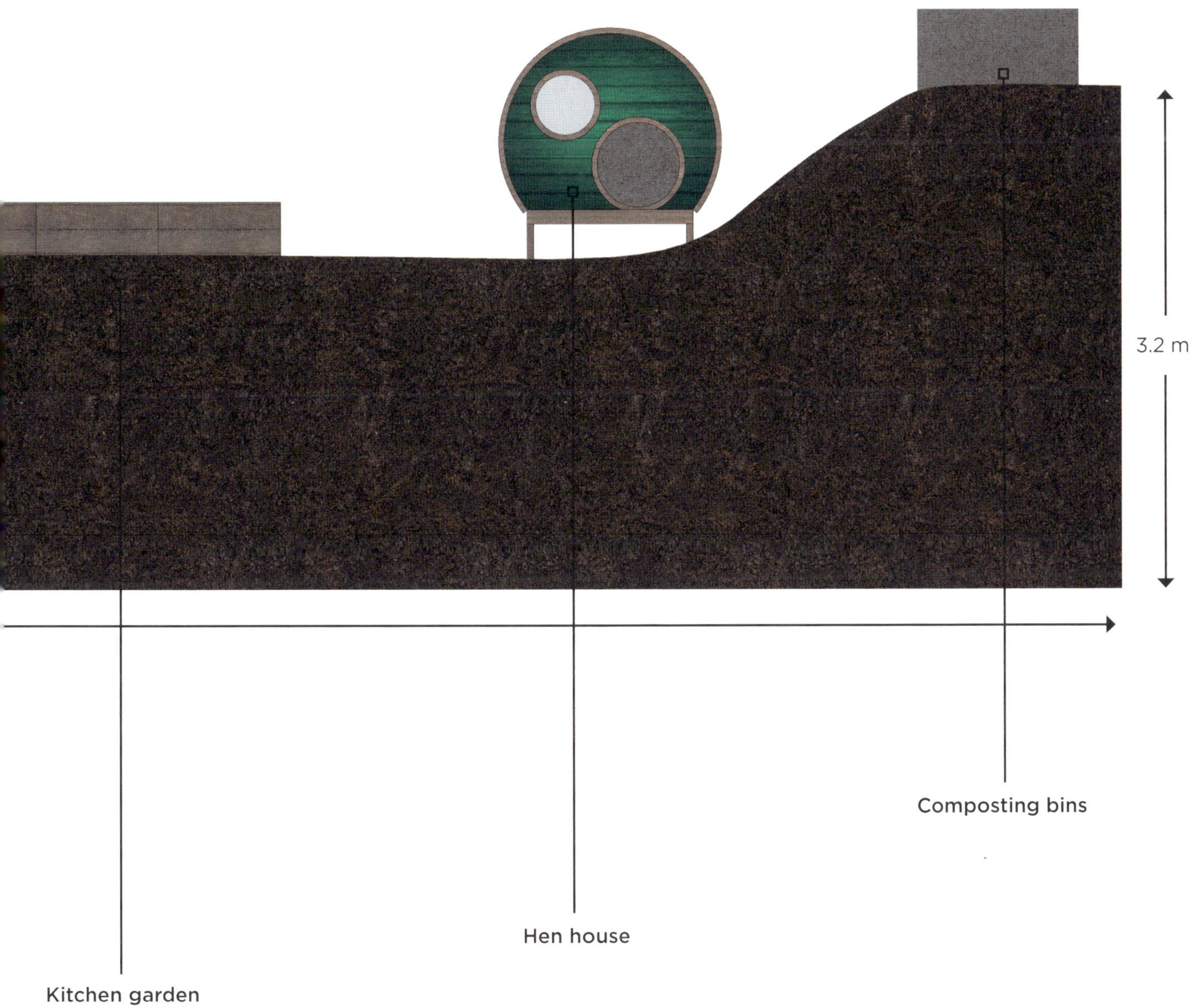

Section

WINGS OF LOVE

For neig

the media, lobbying politicians and p floating community evicted. The most w together with his brother, won the Eurov on the Wings of Love'. But what happen to see the floating community from whe to be living the life that he wants, will Is it because of envy (they don't pay pro are disgraceful) or is it, perhaps, beca of doubt about one's own life? Is it b lifestyle of the boat people present a c

 Was that song maybe more about

ore than a decade, some wealthy
ors on Holmen have been busy engaging
ssuring the administration to have the
al opponent is a singer-songwriter who,
on Song Contest with a song called 'Fly
to those wings? Not only is he not able
he lives, but why is that he, who seems
t let others live the life that they want?
erty taxes) or pity (their living conditions
e the other life gives rise to a smidgen
ause the physical mess and the frugal
llenge to the self-image of the well-off?
elf-love than about the love for others?

неопознанный летающий элемент

[UNIDENTIFIED FLYING OBJECT]

In the 19th century, the Ogu people founded a village called Makoko, on the shore of the Lagos Lagoon, which has since expanded into a huge informal settlement with houses, churches, clinics, schools and shops constructed on stilts in the water, and the people are moving around the labyrinthine waterways between the buildings in canoes.

Buildings are constructed in such a way that they are positioned 4-6 meters above sea level, in order to reduce the effects of flooding and rising sea levels. They are collaboratively built by friends and family members with the use of locally sourced and recycled materials, and oriented so as to reduce solar heating and capture the wind. Water tanks are connected to bore holes, and the water that is used for cleaning, bathing and other purposes is recycled.

The government provides no support - no electricity, water or waste disposal, no healthcare or education. On the contrary, the government actually wants the whole area, which houses several hundred thousands of people, demolished. In 2012, many inhabitants received a notice to move and within a few days thousands of people were rendered homeless before the demolition was halted, after a local community leader was killed.

MAKOKO NOTICE OF EVICTION

"You have continued to occupy and develop shanties and unwholesome structures an the waterfront without authority thereby constituting environmental nuisance, security risks, impediments to economic and gainful utilization af the waterfront such as navigation, entertainment, recreation etc.

"The state government is desirous af restoring the amenity and value af waterfront, protect lives and property, promote legitimate economic activities an the waterfront, restere security, improve water transportation and beautify the Lagos waterfront/coastline to underline the megacity status af Lagos State and has decided to clear all illegal and unauthorised development an its waterfront and water bodies.

"Therefore, notice is hereby given to you to vacate and remove all illegal developments along the Makoko/Iwaya Waterfront within 72 hours af receipt af this notice."

Signed by Akin Tijani on behalf of the Commissioner of Waterfront

POLLUTION

Some neighbor
are defecating
bathe in the water. Unfortunately, the pu
occasionally overflows, making swimmin
have also lodged complaints about the st
However, with the help of volunteers an
the boat people have long since had the
batteries and other polluting contents.
people are destroying nature. To them,
to look at, over there, on the other side
'nice' side, is completely barren. By co
nesting areas for the swans. And if th
defecate in the water, they would

ave spread a rumor that the boat people he water. But of course they aren't. They c sewage system, used by the neighbors, n the harbor dangerous. Some neighbors nded and sunken boats in Erdkehlgraven. he environmental ship, 'Selma of 1914', randed and sunken boats emptied of oil, ally, some neighbors claim that the boat e neighbors, nature is something pretty f the moat, because their own side, the ast, the boat people have constructed neighbors knew how much the swans robably try to have them evicted, too.

"I DIDN'T HAVE A PLACE TO LIVE AND I WAS SICK OF IT ALL. AND IT WAS GOOD TO HAVE A BOAT. IT'S NICE TO HAVE A DOOR."

torken is performing on the road in front of the floating community.

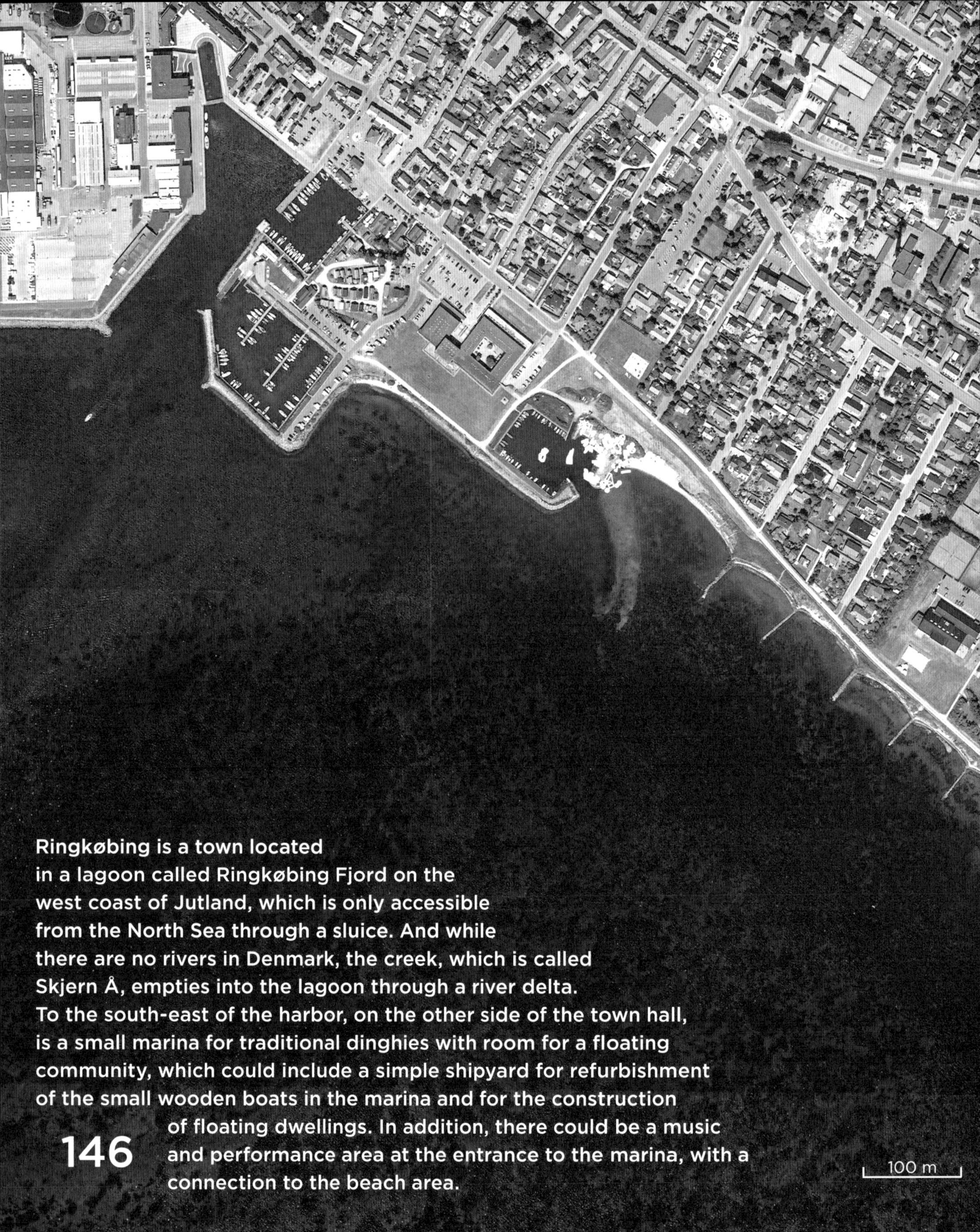

Ringkøbing is a town located in a lagoon called Ringkøbing Fjord on the west coast of Jutland, which is only accessible from the North Sea through a sluice. And while there are no rivers in Denmark, the creek, which is called Skjern Å, empties into the lagoon through a river delta. To the south-east of the harbor, on the other side of the town hall, is a small marina for traditional dinghies with room for a floating community, which could include a simple shipyard for refurbishment of the small wooden boats in the marina and for the construction of floating dwellings. In addition, there could be a music and performance area at the entrance to the marina, with a connection to the beach area.

Ringkøbing
Makeshift shipyard
Bicycle parking
Composting bins
Kitchen garden
Community kitchen
Seating area
Kitchen garden
Music and performance
10 m

"You forced your civilisation upon us and now look where we are: global pandemic, climate crisis, species extinction and, driving it all, widespread spiritual poverty.
In all these years of taking, taking, taking from our lands, you have not had the courage, or the curiosity, or the respect to get to know us.
To understand how we see, and think, and feel, and what we know about life on this Earth."

Nemonte Nenquimo, leader of the Waorani community in the Amazon (2020)
This is my message to the western world – your civilisation is killing life on Earth.
The Guardian, October 12, 2020.

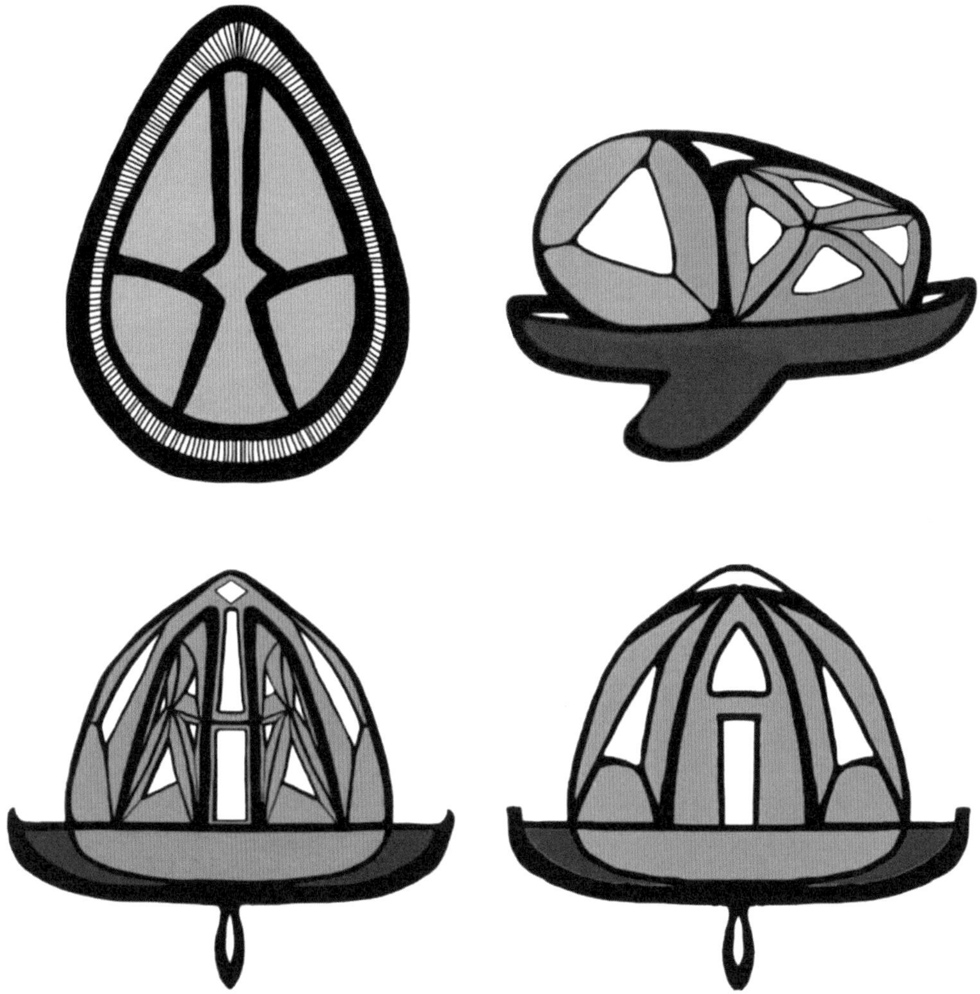

150

"THE CONCEPT 'SEA-HYTTEN' [SEA-COTTAGE] IS PART OF MY VISION FOR A SUSTAINABLE COMMUNITY ON THE WATER, WHICH PROVIDES SOCIETY WITH OPTIONS TO INTEGRATE NEW TECHNOLOGIES WITH ANCIENT MORAL AND ETHICAL IDEALS. HERE, WE COULD FURTHER DEVELOP THE CONCEPT OF DEMOCRACY TO ENCOMPASS EMPATHY: 'EMOCRACY'."

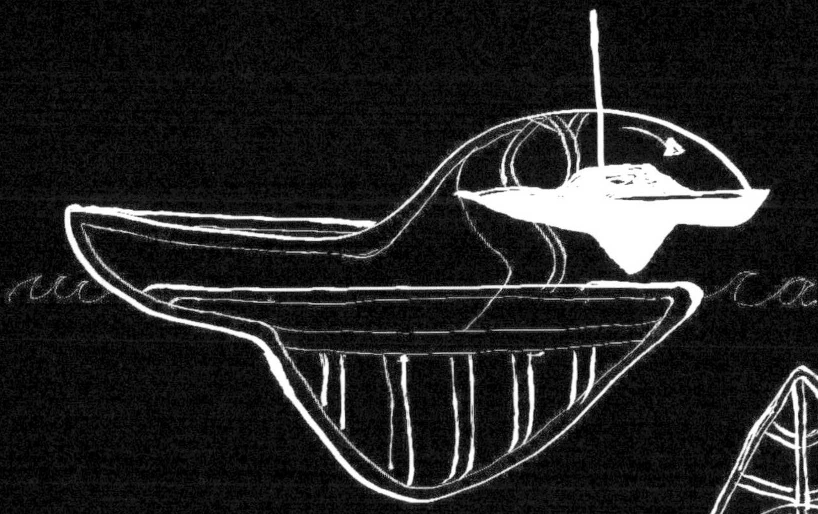

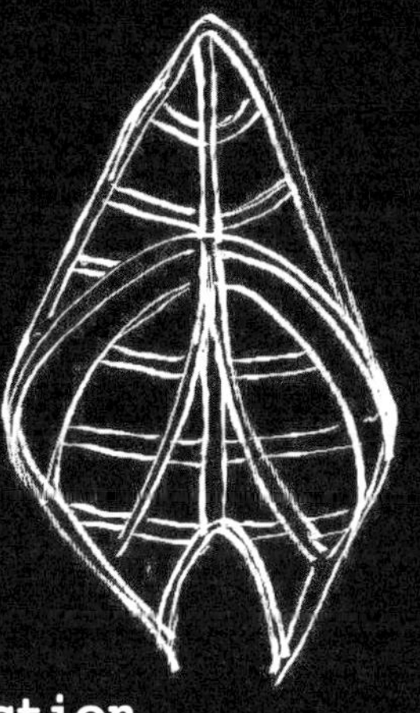

Simple laminated construction.
42 foot deck. Docks with
small craft for mobility,
also serves as dry-dock for
repairs at sea. Optional keel
for deep water stability. To function
as base station for portable marine.

Locations of – and connections to – essential support and resources around the floating community.

Erdkehlgraven
THE ROYAL ACADEMY |
Architecture school
Waste and residual materials
NOMA | Restaurant
Leftover food
COMMUNITY
GARDEN
Seasonal food
NETTO | Supermarket
Leftovers
NEIGHBOR
Drinking water
BADEHUSET
Sauna and shower
DEN GRØNNE GENBRUGSHAL
Builders' merchant
200m

2019 - 2020

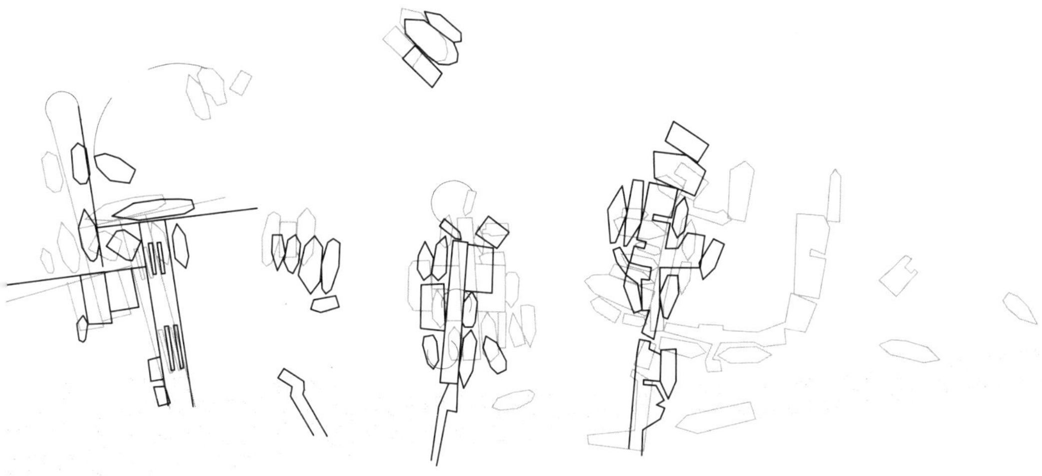

Gabriel

"HOW FAR AWAY DO YOU NEED TO BE IN ORDER TO SEE THE CURVATURE OF THE EARTH?"

LOOKS

Visually, the floating community
– those few neighbors who can a
live. Behind it, the smoking chimneys of two powe
plants produce electricity and heating for the n
the other by incinerating the neighbors' waste.
including an artificial ski slope. Perhaps this ma
that we solve the problems of consumption and p
neat and orderly – some might even say boring –
floating community constitutes a strong sensua
fishing boats, dinghies, self-built rafts and othe
one of the boats. Some people may find it ugly,
It would be like going to Louisiana, the chic mu
and asking to have the paintings by Asger Jorn
and maybe this is just what they should have

they thought it was exciting and now wa

es up very little of the neighbors' field of vision ally see the floating community from where they ants take up a lot more of that visual field. Those nbors, one by burning imported wood chips and n plants were recently wrapped in new designs, them look nicer. But seriously, do we really think ution with a nice wrapping? In comparison to the pearance of the affluent housing on Holmen, the nd colorful contrast, with its worn-out sailboats, ating constructions, and a pirate flag flying over is this really a good reason to have it removed? m of modern art located north of Copenhagen, noved. Then it would be better to stay at home, ne, the people who moved to the city because o have that which is indeed exciting removed.

GARBAGE Due to the currents
the waste that is spill
election posters and other rubbish, ends
for the boat people and their friends, wh
would indeed be 'a garbage dump', whic
by the former Lord Mayor of Copenhager
harassed numerous women at the town h
in a sense, he was right, because the boa
been dumped as garbage by other peopl
garbage than almost any other people in th
get for free, because it would otherwise
And, because modern boats are made o
Instead, they are deposited and those de
community, the discarded boats are

he Inner Harbor of Copenhagen, much of from tour boats, together with politicians' in the reeds in Erdkehlgraven. If it wasn't lean up the area every spring, this place s precisely the expression that was used who was later ousted for allegedly having to describe the floating community. And eople do recycle a lot of material that has on average, Danish people produce more vorld). Their boats, for instance, they often st the owners' money to get rid of them. ass fiber, they are not easy to get rid of. sits are quickly amassing. In the floating en new use, as dwellings for other people.

THE OTHER SIDE OI

Each day during summer, hundrec
close by the floating community in t
the Harbor of Copenhagen. They
and maybe some of them think tha
like a fairy tale than the Little Me
most politicians would probably p
of the glossy image of *Wonderful*
the Welfare State should take car
be taken care of by the State? Ca
they want? Can the State solv

THE GLOSSY IMAGE

if not thousands, of tourists pass
tour boats being navigated around
e clicking away on their cameras,
the floating community looks more
aid, the Royal Palace or Tivoli. But
fer they didn't see this other side
openhagen. Some may think that
of them. But what does it mean to
the State offer people the life that
any of the great questions of life?

MAKING PEOPLE W

are often heard saying that the
judged by how it treats the wea
boat people really weak? On a st
Mette's small boat fell off into th
do? Well, she got up and jumpe
the chimney from about 3 meters
for improvements, but instead o
decided to actively destroy the ho
and the community that they cr

 has decided to make them w

EAK As representatives of the Welfare State, politicians

ystem they represent should be

st people in society. But are the

ming winter night, the chimney of

roaring sea. Then, what does one

nto the cold water and retrieved

epth. Yes, there is definitely room

helping with this, the system has

es people have made themselves,

ted and, in this way, the system

k. And dependent on the system.

cultures, in particular those related to the s
and counter-cultures, as well as the c
movements. Regrettably, all these people,
sustainable, are being suppressed and c
whose lifestyles are anything but sustaina
only accumulate enough resources to me
frequently move around, they make clos
Or rather, they didn't make any lasting im
of living has largely been rendered impos
interests of colonial powers and the settle
replaced their societies – most of which
sense of giving every member a voice in t
sharing the available resources amo

The floating community shares many
similarities with indigenous nomadic
, but also with contemporary autonomous
it-yourself, simple-living and tiny-house
ose lifestyles are often very close to being
acized by the majority of the population,
e. Most nomadic sea people, for instance,
their immediate needs, and because they
o no lasting impact on the environment.
ct on the environment, because their way
ble by the administrative and commercial
ocal population, who have supplanted and
em to have been truly democratic, in the
decision-making and in the sense of fairly
all members – with modern 'democracies'.

"MY BOAT WAS HALF SUNKEN IN THE HARBOR. SOME PIRATES HELPED ME TAKE THE ENGINE FROM ANOTHER BOAT AND I PUMPED THE WATER OUT OF IT AND MADE IT LIVABLE. WE PAINTED THE BOTTOM OF THE BOAT AND I RENAMED IT JACK SKULL (INSTEAD OF MY WAY). THEN I MOVED IT OVER NEXT TO STORKEN'S BOAT."

MY WAY

Various people worked on a mandala, that ended up in the ceiling of the boat Caroline.

"... it is not love and not even sympathy upon which Society is based in mankind. It is the conscience – be it only at the stage of an instinct – of human solidarity. It is the unconscious recognition of the force that is borrowed by each person from the practice of mutual aid; of the close dependency of every one's happiness upon the happiness of all; and of the sense of justice, or equity, which brings the individual to consider the rights of every other individual as equal to his own."

Peter Kropotkin (1902)
Mutual Aid – A Factor in Evolution. McClure, Philips & Company, pp. 4-5.

Organic waste is composted for two years inside a container made of recycled wood, which is properly ventilated and placed in a semi-shaded location, so as to avoid the compost drying out or rotting.

Compost is extracted from the bottom of the container where it is furthest developed.

Composting

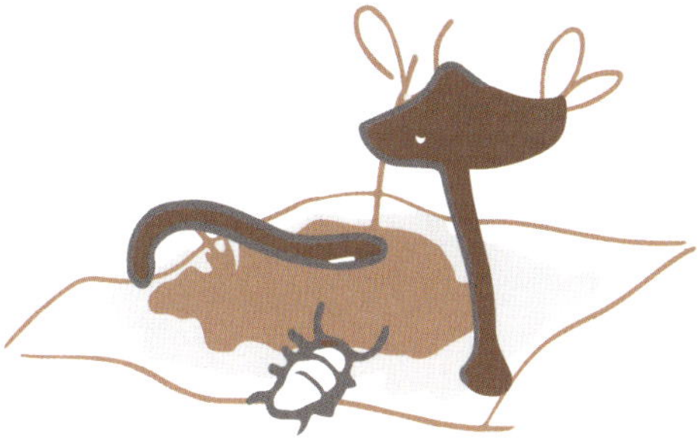

The compost develops a combination of microorganisms, fungus and nutrients, which is then used to enrich the soil and thus stimulate the growth of plants – for instance, in the kitchen garden – and to retain water and avoid soil erosion.

GETTING RID OF THEM

Copenhagen, decided to spend 26 million k and the rest from the state) to have the floa which is invested with the authority over the injunctions to the people who identified the same time, a company was hired to clean up not have an identified owner. This caused a nu indicted and subsequently convicted, in both complied with the injunctions. According to la the Coastal Authority is entitled to appeal anc in such a bureaucratic language, in the injur none of the meetings with the authorities was for making an appeal had, by then, been e

injunctions were indeed legitimate, only

ı 2019, the center-right government of Denmark, ɔgether with the center-left Lord Mayor of er, in total (10 million from the municipality ƺ community removed. The Coastal Authority, :er in Erdkehlgraven, then began handing out elves as owners of floating dwellings. At the e area, and to dispose of everything that did er of people to flee the area while others were e lower and the higher courts, for not having a person who has been given an injunction by have their case tried, but this was formulated ons, that no one could understand it, and in is ever mentioned. And because the deadline eded, neither court considered whether the ether people had complied with them or not.

The Moken used to roam the Andaman Sea and the Mergui Archipelago in their kabangs – small handcrafted boats, which included a kitchen, a bedroom and a living-area for one family. They would usually move around in groups of up to ten boats.

During the rainy monsoon season (May – December), they would build temporary bamboo huts on stilts along the shore. When a new family was formed, the whole group would come together to build them a kabang.

They lived off the fauna and flora of the sea – fishing with spears and free diving for sea cucumbers and shellfish. It is said that their children learned to swim before they could walk.

Because their dwellings were mobile, conflicts could be settled simply by moving away and epidemics could be avoided by leaving infected people with provisions in one place, hoping they would recuperate, while the others moved on.

By fleeing to higher ground or heading out to deeper parts of the sea, most of the Moken escaped the devastating tsunami that hit this part of the world on December 26, 2004, and left hundreds of thousands of people dead along the coasts of Indonesia, Sri Lanka, Thailand and other countries. Even though they had managed to save themselves by picking up on the early warning signs of the tsunami, they were nevertheless evacuated by the authorities and aid organizations in the aftermath and resettled, far away from their natural habitat, the sea.

Moreover, their lifestyle has been rendered impossible by offshore oil drilling and industrial fishing, mass tourism and development, abuses by the military and harassment from authorities. And, bizarrely enough, also by ecological conservation efforts, which prohibit the Moken people from felling trees to make their kabangs and from collecting sea products that they would traditionally trade for rice and other necessities.

Being stateless, they are also prevented from moving around freely on the sea, as well as being evicted from land – because they don't own it: they don't want to own it, and they're not allowed to.

They find it hard to adapt to modern life because they believe natural resources, including land, food and material goods, should not be owned by anyone but should be shared among everyone.

"NINA, THE BLACK HAIRED BEAUTY."

Around the boat Caroline, there were the boats of Nassir, Joâo and Monika. Together, they formed the Nameless Island.

"CAROLINE WAS CREATED AS A HEALING SPOT, FOR THE PEOPLE WHOSE HEART WAS SCREAMING, ABOUT HOW THE WORLD IS WORKING, WITH ABUSE AND HUNGER ALL AROUND THE WORLD. CAROLINE WAS, AS WELL, A SPOT THAT WAS SAFE FROM PATHOLOGICAL BEHAVIORS ON LAND."

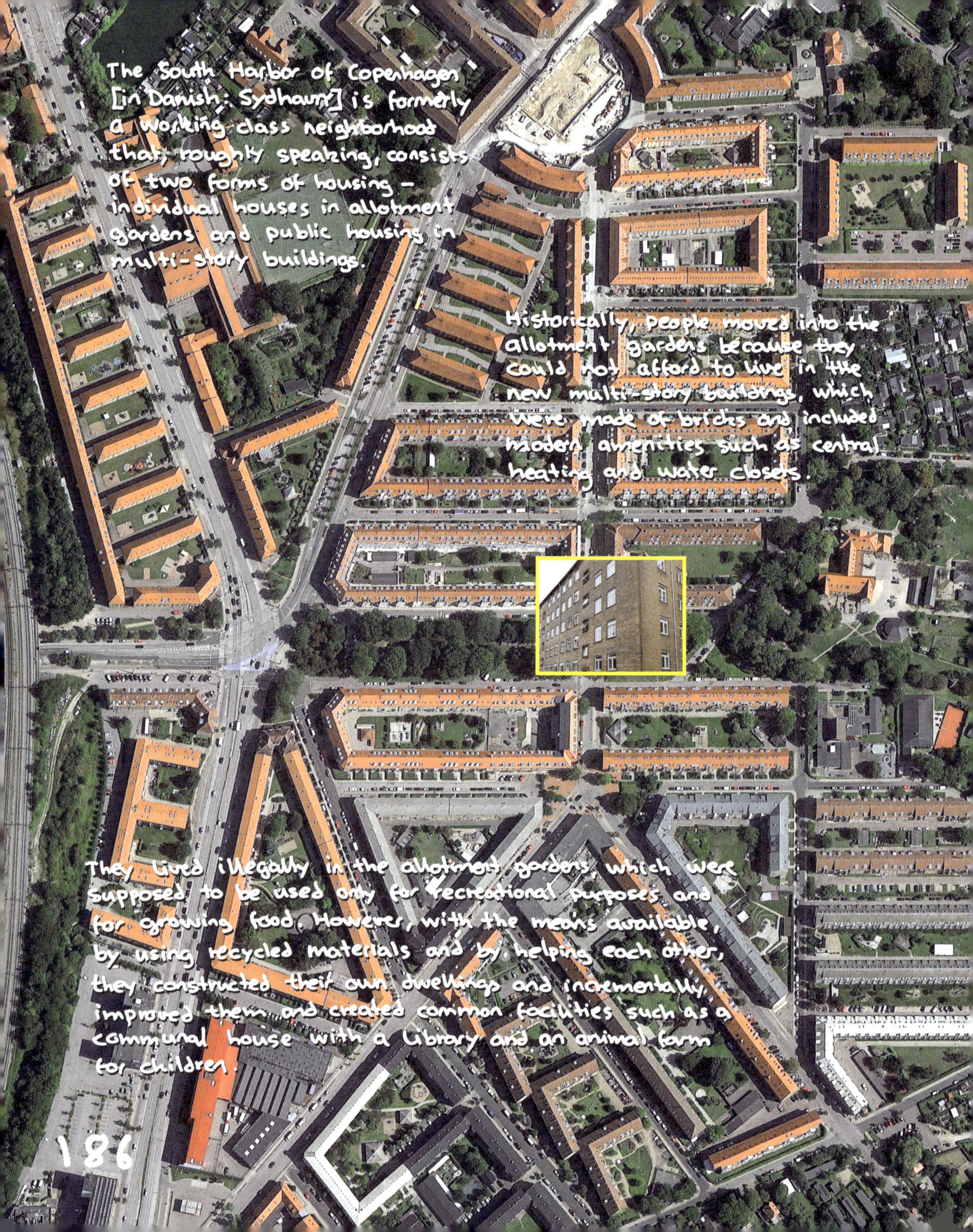

The South Harbor of Copenhagen [in Danish: Sydhavn] is formerly a working-class neighborhood that, roughly speaking, consists of two forms of housing – individual houses in allotment gardens and public housing in multi-story buildings.

Historically, people moved into the allotment gardens because they could not afford to live in the new multi-story buildings, which were made of bricks and included modern amenities such as central heating and water closets.

They lived illegally in the allotment gardens, which were supposed to be used only for recreational purposes and for growing food. However, with the means available, by using recycled materials and by helping each other, they constructed their own dwellings and incrementally improved them and created common facilities such as a communal house with a library and an animal farm for children.

Sydhavn

Eventually, taking up residence in most of the allotment gardens was legalized and gradually, the status of the two forms of housing changed: What was once perceived as slum, the allotment gardens, now became more and more attractive, and correspondingly expensive. So expensive, indeed, that the people who once moved in there because they had no other option, would be unable to do so today. Instead, they would probably have to settle for a subsidized apartment in the multi-story buildings. Because even though those buildings were planned and designed with the best intentions and facilities that were available then, it was not possible for the residents to make any real improvements or create anything new. Therefore the public housing, which was once perceived as a modern form of living, has since decayed into a contemporary form of slum.

"... the nomad is not necessarily someone who moves around: some journeys take place in the same place, they're journeys in intensity, and even historically speaking, nomads don't move around like migrants. On the contrary, nomads are motionless, and the nomadic adventure begins when they seek to stay in the same place by escaping the codes."

Gilles Deleuze (2004 [1972])
Desert Islands and Other Texts 1953-1974.
Los Angeles, Semiotext(e), pp. 259-260.

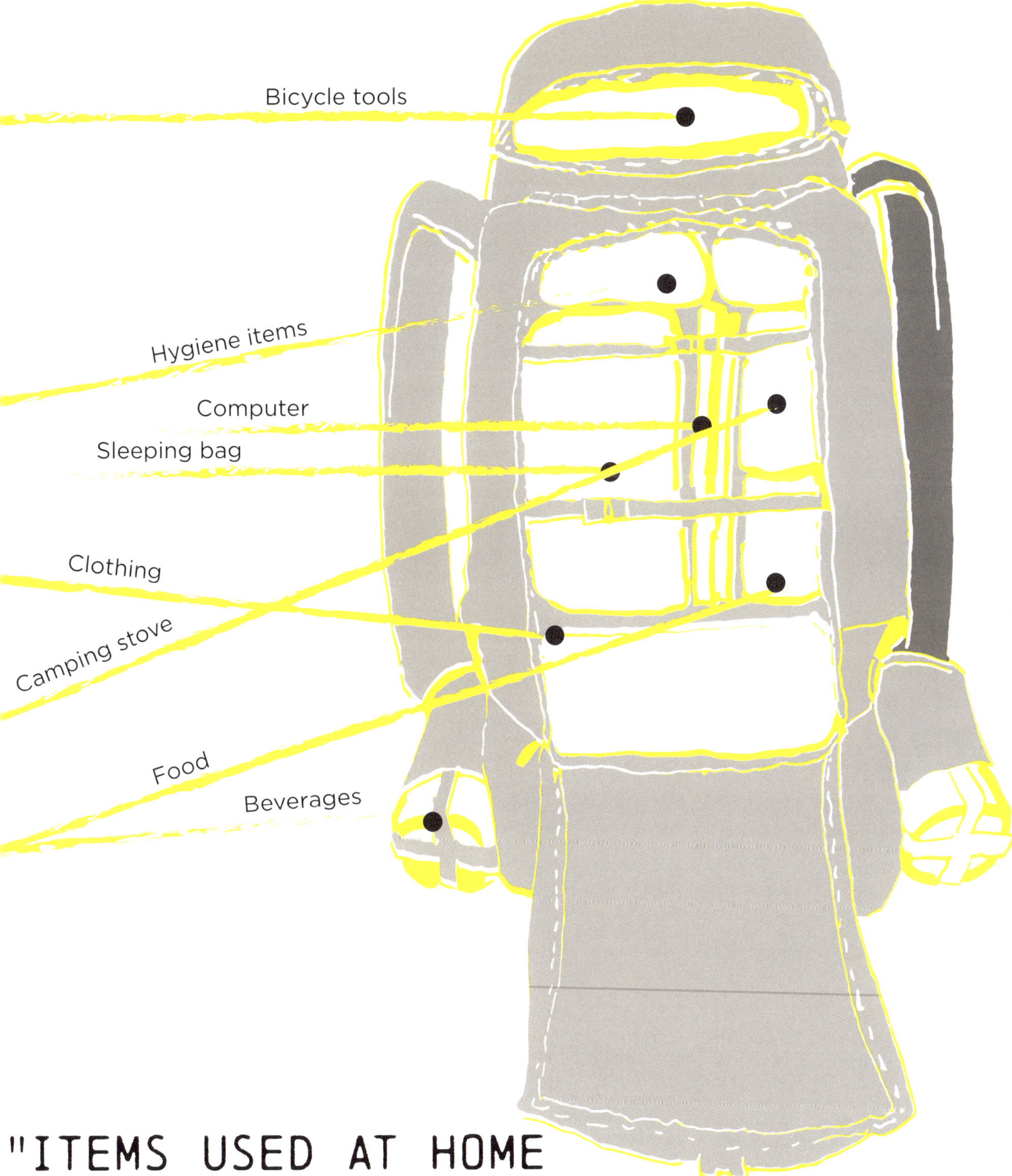

"ITEMS USED AT HOME
SHOULD BE SMALL AND PORTABLE, SO THAT
I CAN ALSO USE THEM WHEN TRAVELING."

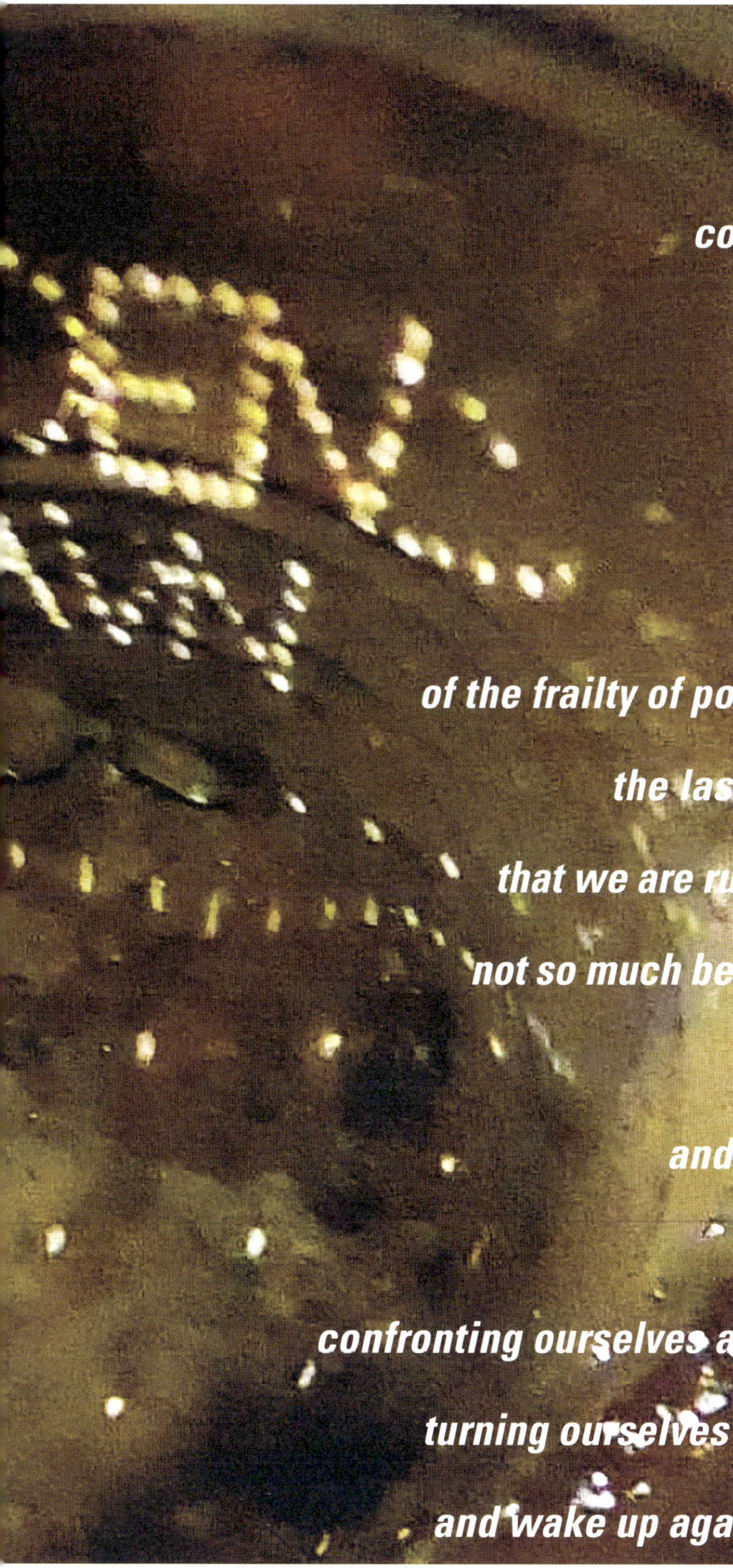

HARBOR OF PEACE

You wonderful sea cow

composed of quirky characters

ll of alluring frequencies

effects of drunken people

full of frustration, revolution,

demonstration

of the frailty of power and the power of frailty

the last will be first to understand

that we are running in the wrong direction

not so much because the direction is wrong

but because we run

and then we stop, wake up from

the neo-dream

confronting ourselves and not being able to stand it

turning ourselves around ourselves till we fall

and wake up again . . . in the Harbor of Peace.

A QUESTIONABLE DECISION A fl

This decision was largely based on a statement provide City of Copenhagen, which had been approved by the p According to this statement, the floating community future use of the area". Furthermore, it "appears, both therefore "does not contribute to the beautification of However, with regard to the future use of the area, the n Erdkehlgraven but not the water) specifies that the area wh public recreational purposes, which would, it must be assum The issue of appearance is, of course, both a relative and a of some of the neighbors) as the basis for its response mak just as well be argued that worn-out vessels that are being r But what really makes the statement dubious is that: " from toilets and other sources is being discharged into the municipality had already commissioned a company statement was drafted, the municipality had received the

It was not until almost a year later that a test – municipality's own expert, the bacteria (E.coli) f

dy in 2014, the Coastal Authority had declared the
ng community to be illegal and to be subject to removal.
y the Technical and Environmental Administration of the
cal committee of the administration, with 6 votes to 5.
es not comply with the municipality's plans for the
land and at sea, very disorderly and full of waste" and
confer any amenity value to the area. On the contrary."
cipality's own local plan (which includes the land around
the floating community garden is located is to be used for
include a community kitchen garden, kitchen and campfire.
ective matter. Thus, to use a particular ideal of beauty (that
he municipality appear rather prejudiced. Because it might
ed and the biotopes that emerge around them are beautiful.
municipality believes that, in all probability, wastewater
area." In fact, as a result of the neighbors' complaints,
test and analyze the water quality, and by the time the
t results, none of which showed any signs of wastewater.
ber 29 – finally showed such signs, but according to the
d in this test probably had its source in the waterfowl.

Elements are incrementally added and modified.
196

Light materials and structures are used.

The Tankas used to live on junks - sailing ships that derived their design from ancient Austronesian ships and featured several innovations that were later adapted by the Europeans, including the transverse hull separation of watertight compartments divided by bulkheads.

They don't have their own language or a distinctive culture except for some folk songs and a few other cultural elements, such as hats

They are probably descendants of local aboriginals, rebels and refugees who have fled to the sea throughout Chinese history, and foreigners. Tanka is a derogatory term in the Cantonese language and the Tanka people have historically been treated as outcasts. During certain periods, they were forbidden to settle on shore and to intermarry with the local people. Living on boats was, however, also a cheaper option than having to pay for land and a house on land and being subject to abuse from the land-based population; living on boats also provided some form of safety.

Having no land and only few possessions to pass on to new generations may explain why the Tankas had no social organization based on lineages or kinship.

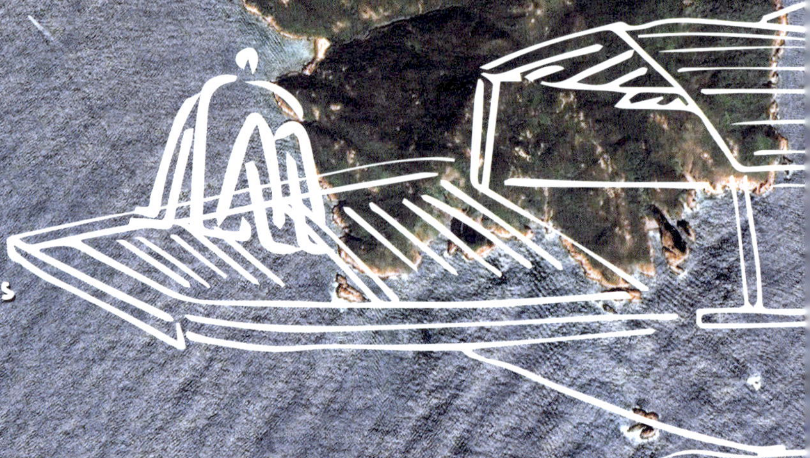

The Tankas were fishermen but would also ferry cargo and passengers across the waters. Fishing may have contributed significantly to a healthy diet among the local population but this particular occupation – killing living animals – may also have put them at odds with the prevailing Buddhist religion of the area, thus contributing to their being ostracized from society – in addition to their status as being simply 'different' and therefore being considered 'barbarians'.

By the 1970s, the number of boat people living along the coast of southeast China was reported to be at least 3 million, but today their way of life has largely disappeared.

makes homeless people even more
surprising to see the eagerness with
speedy handling of the cases against
since these cases could potentially res
rendered homeless. While the munici
take away people's homes, there was ze
who were about to be made homeles
been set aside for municipality workers
about what their opportunities might b
are severely limited. Anyway, the boat

 want to keep their homes and the

MELESS

Given the corona pandemic, which

nerable than they usually are, it was

ich the authorities were pushing for a

e boat people, in the juridical system,

t in a number of the boat people being

ity had allocated 10 million kroner to

kroner in the budget to help the people

although 400.000 kroner had indeed

o reach out and inform the boat people

as homeless people, opportunities that

eople are not asking for help. They just

mmunity that they themselves created.

Collecting bottles on the street has been Kami's main income for a long time. She can make around 150 DKK - 200 DKK on a good day.

Kami recently started selling postcards, T-shirts, and a magazine about homelessness, which gives a better income than collecting bottles.

A raft that was built in the Do-it-yourself-harbor in Sydhavn ended up i Erdkehlgraven, where a shack that once had been used for various things on land was placed on top of it. For some years hence, it functione as a floating café, selling pancakes and coffee, and later on, it wa

ısed as a shared kitchen by the people living in the Floating Quarter. ˈtorken slept there for two years with his girlfriend. The nameplate, Havørnen' [The Sea Eagle], he recovered from a half-sunken ship.

BEING ANCHORED

is technically part of the Danish coas
law, to lie anchored in the same place
people were to occasionally leave Erd
period of time, they could be there, le
they are afraid of leaving since they f
the local bridge association, which co
itself controlled by the neighbors, won
be against the law, that stipulates tha
But can the boat people count on t
case? Would the neighbors even

Despite Erdkehlgraven being located
in the Inner Harbor of Copenhagen, it
ıe, where one is allowed, according to
at least a few months. Thus, if the boat
ılgraven, and go somewhere else for a
lly. But where would they go? Besides,
r, based on practical experience, that
ols the access to Erdkehlgraven and is
et them in again. This, of course, would
ıere has to be free access for anyone.
authorities upholding the law in that
e if they were fined a meager penalty?

The Tlingit, Haida, Tsimshian and Kwakiutl are tribes that used to live on the sea and rivers along the Pasific Northwest Coast. They used to fish for their livelihood, rather than hunting or farming, while skillfully navigating their canoes around the islands.
They were frequently on the move to visit other tribes, clans or families, especially during winter when social life was very intense. There were feasts related to initiation, marriage, funeral and many other rituals. These feasts often involved music, dancing, singing, storytelling, games and potlatches, in which gifts were given to enhance relations, determine positions and settle disputes. Potlatching evened out material inequality but gifts were sometimes rejected
This practice was banned in 1884 because it was found to contradict the 'civilized' values of production and accumulation that had become part and parcel of modern society.

During these rituals, the masks that were worn could be opened to symbolize the transformation of animals into ancestors and mythical beings. And the totem poles that were erected were carved with symbols of family ties, stories and beliefs, sometimes commemorating someone and sometimes shaming someone.

Many aspects of these spiritual rituals were, however, condemned by missionaries and were prohibited by law until the second half of the twentieth century.

"WHEN I STUDIED THEATER, WE USED MASKS, AS IN OLDEN DAYS, TO HIDE OUR FACES AND EMPHASIZE OUR BODILY MOVEMENTS. LATER, WHEN I BEGAN SCULPTING IN STONE, I CREATED AN INFINITE SERIES OF FACES CALLED 'SLEEPING GODS'."

Stephen James Back

"[To the Trobiand people of Papua New Guinea,] the large abalone shells, the shields covered with them, the decorated blankets with faces, eyes, and animal and human figures embroidered and woven into them, are all personalities.

The houses and decorated beams are themselves beings. Everything speaks – roof, fire, carvings and paintings; for the magical house is built not only by the chief and his people and those of the opposing phratry but also by the gods and ancestors; spirits and young initiates are welcomed and cast out by the house in person."

Marcel Mauss (1967 [1923])
The Gift. Ian Gunnison (transl.). New York, W. W. Norton & Company, p. 43.

The World is at my feet, no
questions, no place to sleep.
As silence breaks, you just wanna
look away.
Adventure, you unknower.
Let's sail out into the lost night.
While distance between us is all too
long.
Stranger be my comfort zone.
When the smoke is gone and the
dope all away,
when them 3000 Didi-Dada tunes
ring out in my brain.
There's only just something left.
What something,
is left to be found out.
First of all, we have to trust it.
For this, we need to trust our selves.
Our selves being all of us.

Kasper Hedegreen

Kasper is playing guitar with his band at a party in the community garden.

nature. They are affected when the wind is b
freezing, when the sea is roaring, when the
when the flowers are blooming ... And they tr
By contrast, settled people employ scie
manipulate and discipline nature – both the
While the vast majority of the settled p
represented by the state, is fighting nature, u
being represented by anyone but themselve
of human nature that doesn't want to be exp
When modern people hear this cry for free
disbelief. Because rather than living in a
resemble that of robots. (It's no wond

going to take away from us what we

TS Like other nomadic sea-cultures, the boat people live in close contact with

ving, when the rain is falling, when the soil is
s warming, when the waterfowl are nesting,
adapt, using both creativity and spirituality.
e and technology to exploit and control,
st outer nature and the inner human nature.
lation, their institutions and companies,
ettled people, like the boat people, not really
are fighting for nature, including that aspect
ed or controlled, manipulated or disciplined.
m, they tend to regard it with mistrust and
ee and natural way, our lives increasingly
that we are afraid that real robots are
old to be the meaning of our life – work.)

Örn created a windmill, using a plastic basket, a metal structure, old records, and a drilling machine.

BORUP

The Marsh Arabs used to live in the marshlands of southern Iraq and on the border between Iraq and Iran. They constructed artificial islands of reeds, on which they erected small dwellings, also made of reeds, one part being reserved for people and another part for animals.

In order to navigate their way through the marshlands and rivers, they used diffirent types of slender and flat-bottomed canoes – made of reeds, wood and woven mats – including the large tarada, with its long curved prow, for warfare and hunting, and the smaller mashoof canoe, dating back at least 7.000 years, for fishing, gathering reeds, transportation and for driving their water buffaloes through the reeds and grasses to create channels in the marshes.

For centuries, the marshland has been used by slaves and rebels as a hiding place. During Saddam Hussein's reign, the marshes were drained and villages were burned and bombed in order to force insurgent Marsh Arabs out and make them give up their way of life.

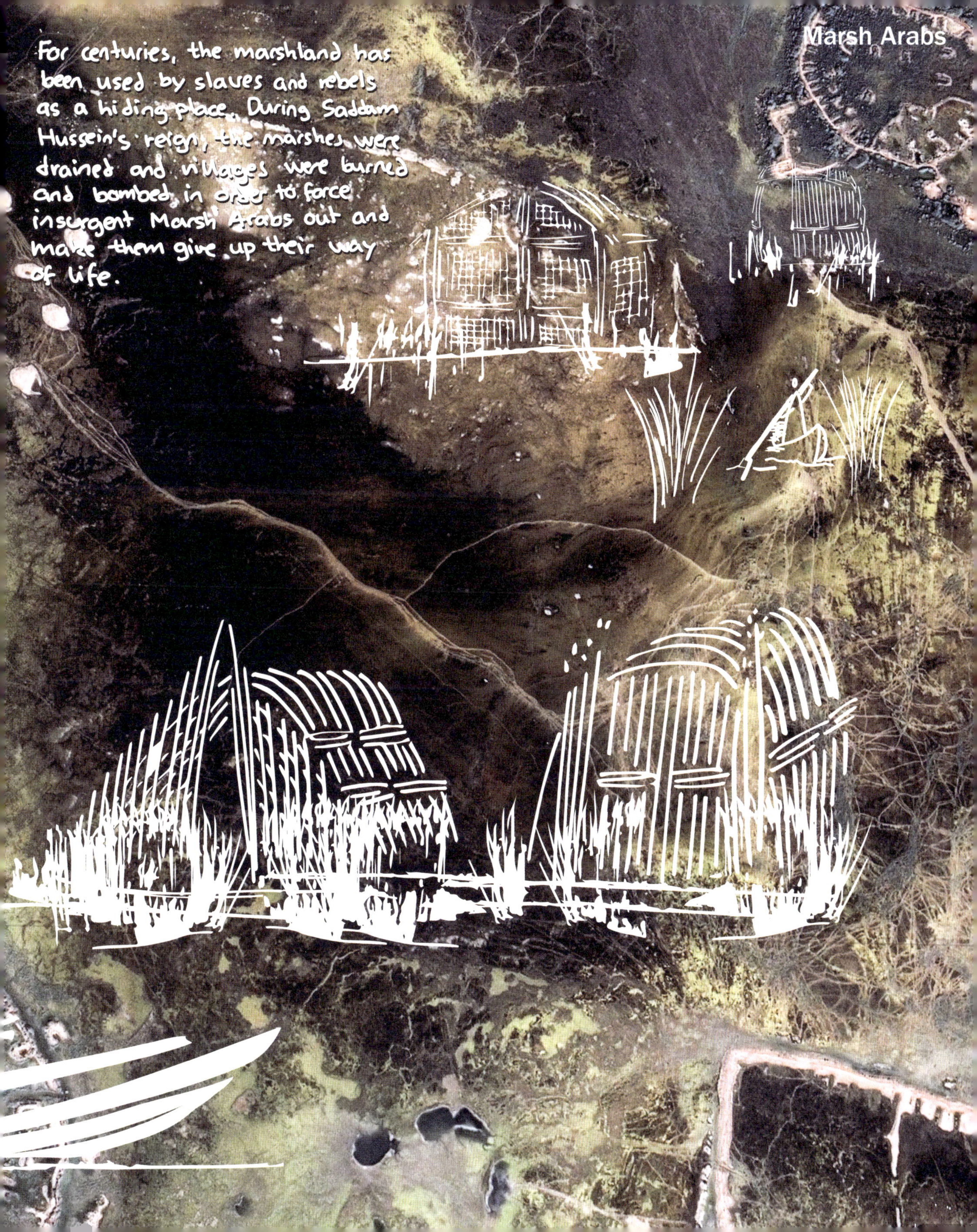

"IT WAS THE REALIZATION OF AN IDEAL
SURVIVING BY YOURSELF AND NOT HARMI

"IC ATTITUDE TO LIFE:
OTHERS."

"IT WAS A DREAM OR NIGHTMARE –
DEPENDING ON THE MOMENT.
MOSTLY A DREAM".

" ... the question of what kind of city we want cannot be divorced from the question of what kind of people we want to be, what kinds of social relations we seek, what relations to nature we cherish, what style of daily life we desire, what kinds of technologies we deem appropriate, what aesthetic values we hold. The right to the city is, therefore, far more than a right of individual access to the resources that the city embodies: it is a right to change ourselves by changing the city more after our heart's desire. It is, moreover, a collective rather than an individual right since changing the city inevitably depends upon the exercise of a collective power over the processes of urbanization."

David Harvey (2003)
The Right to the City.
International Journal of Urban and Regional Research, 27, 4, 939-941. DOI: 10.1111/j.0309-317.2003.00492.x.

"WE DON'T HAVE A HARBOR-MASTE

O I JUST GAVE THEM MY ADDRESS"

"There is nothing that man fears more than the touch of the unknown.
He wants to see what is reaching towards him, and to be able to recognize or at least classify it.
Man always tends to avoid physical contact with anything strange.
In the dark, the fear of an unexpected touch can mount to panic.
Even clothes give insufficient security: it is easy to tear them and pierce through to the naked, smooth, defenceless flesh of the victim.
All the distances which men create round themselves are dictated by this fear.
They shut themselves in houses which no-one may enter, and only there feel some measure of security."

Elias Canetti, (1981 [1962]) *Crowds and Power.* New York, Continuum, p. 15.

FORGYLDTE MOMENTER
ENEVÆLDIGE VÆLTER

HVORDAN VILLE NOGEN MENNESKER, KUNNE SKABE
SIG EN TILVÆRELSE I DET KAPITALISTISKE
SAMFUND UDEN AT LÆGGE ET LÅG PÅ SIG SELV?
UDEN AT KOMME LÆNGERE VÆK FRA SIN KERNE?

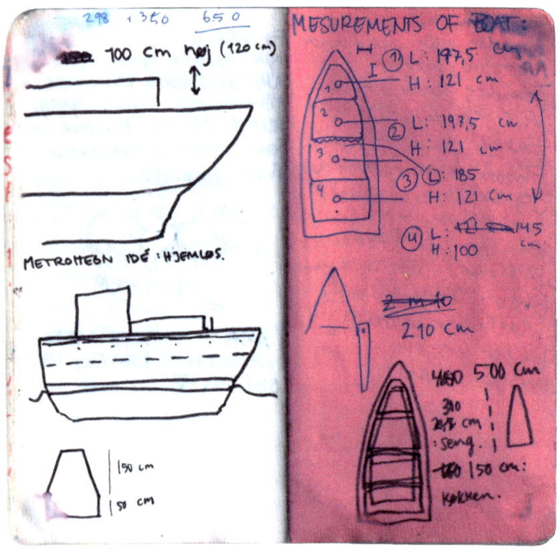
298 + 350 650
100 cm høj (120 cm)
METROHEGN IDÉ: HJEMLØS.
150 cm
150 cm
MESUREMENTS OF BOAT:
① L: 197,5 cm
H: 121 cm
② L: 197,5 cm
H: 121 cm
③ L: 185
H: 121 cm
④ L: 145 cm
H: 100
210 cm
500 cm
seng.
150 cm: køkken.

ZORRITA

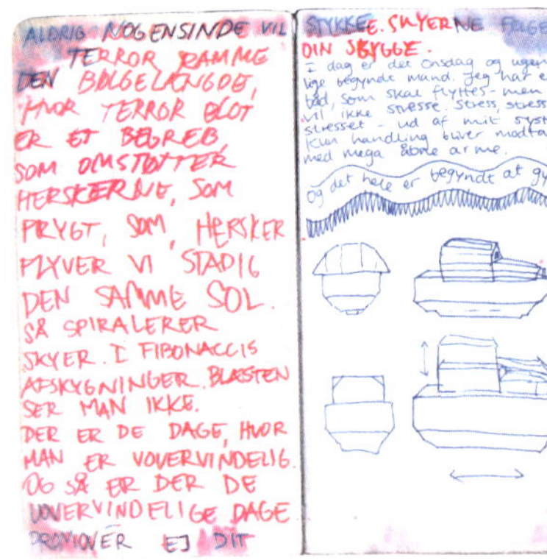
ALDRIG NOGENSINDE VIL TERROR RAMME DEN BØLGELÆNGDE, HVOR TERROR BLOT ER ET BEGREB, SOM OMSTØTTER HERSKERNE, SOM FRYGT, SOM, HERSKER FLYVER VI STADIG DEN SAMME SOL. SÅ SPIRALERER SKYER. I FIBONACCIS AFSKYGNINGER. BLÆSTEN SER MAN IKKE. DER ER DE DAGE, HVOR MAN ER UOVERVINDELIG. OG SÅ ER DER DE UOVERVINDELIGE DAGE
PROVIVER EJ DIT
STYKKE. SKYERNE FØLGE DIN SKYGGE.
I dag er det onsdag og ugen lige begyndt mand. Jeg har en båd, som skal flyttes - men vil ikke stresse. Stress, stresse stresset - ud af mit system. Kun handling bliver modtaget med mega åbne arme.
og det hele er begyndt at gy

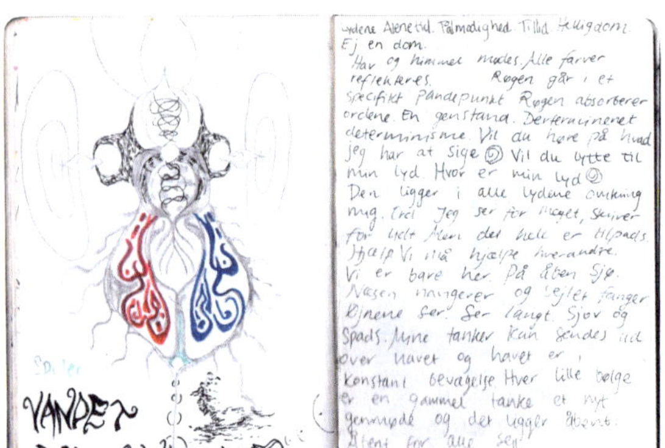
VANDET ER LYDENE
lydene Alene tid. Tålmodighed. Tillid. Helligdom. Ej en dom.
Hav og himmel mødes. Alle farver reflekteres. Røgen går i et specifikt pandepunkt. Røgen absorberer ordene. En genstand. Determineret determinisme. Vil du høre på hvad jeg har at sige Vil du lytte til min lyd. Hvor er min lyd
Den ligger i alle lydene omkring mig. Jeg ser for meget, skriver for lidt. Men det hele er tilpads. Hjælp. Vi må hjælpe hinandre. Vi er bare her. På åben sjø. Næsen navigerer og sejlet fanger. Øjnene ser. Ser langt. Sjov og spøg. Mine tanker kan sendes ud over havet og havet er i konstant bevægelse. Hver lille bølge er en gammel tanke et nyt gennemødt og det ligger åbent. Åbent for alle sejl

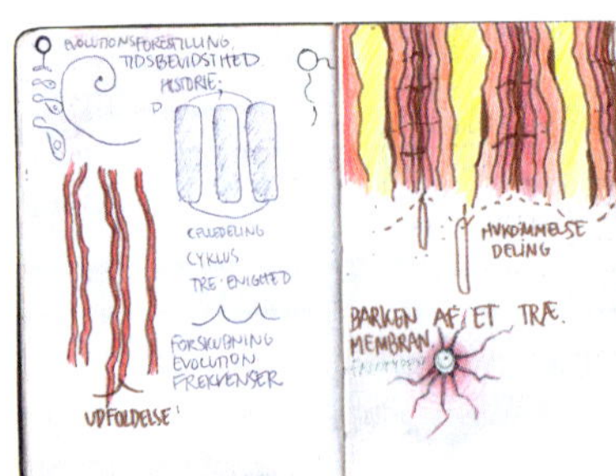
EVOLUTIONSFORESTILLING,
TIDSBEVIDSTHED
HISTORIE:
CELLEDELING
CYKLUS
TRE·ENIGHED
FORSKUBNING
EVOLUTION
FREKVENSER
UDFOLDELSE!
UFULDKOMMENHED
DELING
BARKEN AF ET TRÆ.
MEMBRAN.

Sira Hentze

DU ER BARE MEGA DEJLIG
SKAL VI KOMME SAMMEN?
ja nej måske
BÅD-NOTER.
Den første ejer af båden har lige været forbi.
Den oprindelige ejer.
Information:
Båden tilhører ikke
Jeg skal kunne komme til, i bunden af båden, rense, fra midten af.
Politiet har været der flere gange og spurgt til ejerne af båden, men det er ikke noget alvorligt

BÅNDOPTAGER
FLERE LAG SOM I FLERE FREKVEN SER INTAKT ELLER DE TAKT MED ALLE TAKTERNE DU KAN ALDRIG UNDGÅ AT VÆRE PÅVIRKET AF NOGET NÅR KROP OG SIND IKKE ER CONNECTED = DØD

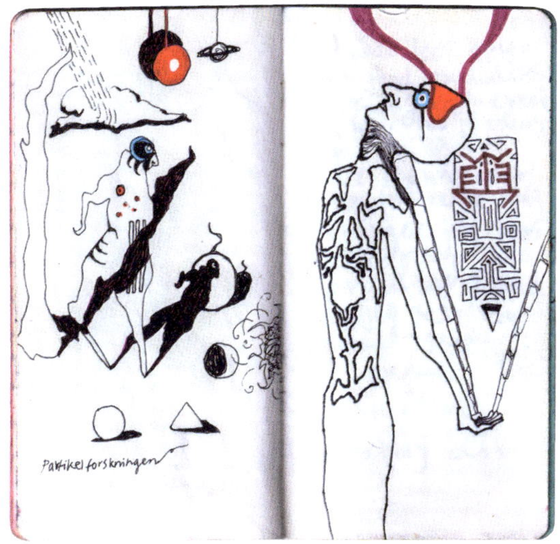

Partikel forskningen

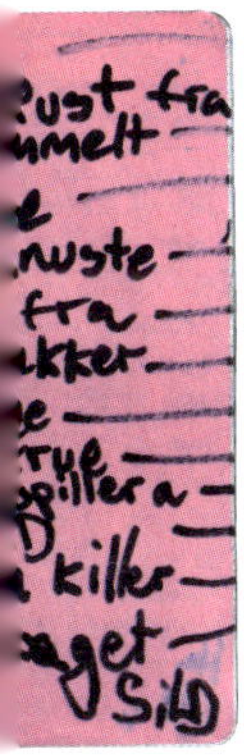

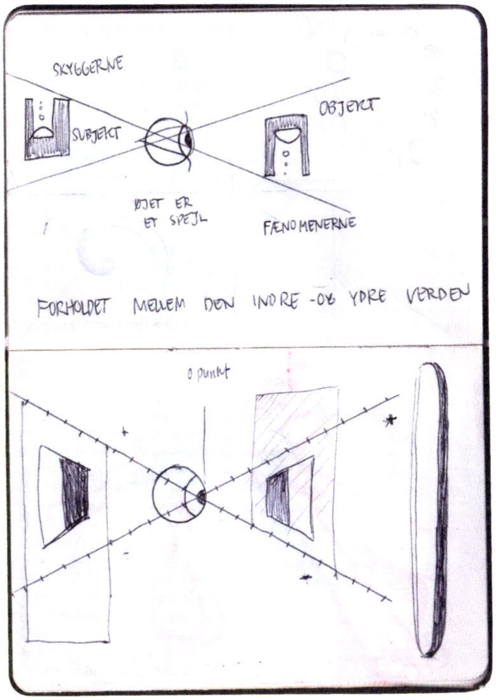

SKYGGERNE
SUBJEKT
OBJEKT
ØJET ER ET SPEJL
FÆNOMENERNE
FORHOLDET MELLEM DEN INDRE- OG YDRE VERDEN
0 punkt

HISTORIEN OM DE RUNDE STEN, REFLEKTERER VINDENS FARVER.

Samme stemning Sang
Blå og gul.
Udefineret forhold.
Lyden af gul.
KIRKE KOR
Fisk hopper

THE INUIT AND THE NO

and the culture of the settled – and what such a d
change (and other challenges) – is the story of the
Greenland during the latest period of climate cha
consisted of provisional skin-made tents and pea
caught at sea (during winter), the game they hunte
the 'privileged' Norsemen had livestock, such as
houses to live in, all year. But as the climate got co
became increasingly dependent on the seals they
houses – and churches! – were built of stone me
not follow, as the Inuit were able to do, the chang
therefore, became more and more monotonous, w
– or, in other ways, to having disappeared. The
insofar as nature was more or less stable a
way of life and adapting to a persistent d

SEMEN An illustrative example of the difference between nomadic culture

erence might imply in terms of resilience to climate t and the Norsemen, who simultaneously inhabited e, the Little Ice Age. While the homes of the Inuit uts, and their food was comprised of the fish they and the berries they found inland (during summer), tle, sheep and goats, to feed on, and solid stone r, the livestock began to die out and the Norsemen ld catch near their settlements. The fact that their that they could not move their homes and could patterns of flora and fauna. The Norsemen's diet, n may have contributed to them leaving Greenland semen were skilled in exploiting nature, but only predictable. They were incapable of changing their rioration of the climate over several generations.

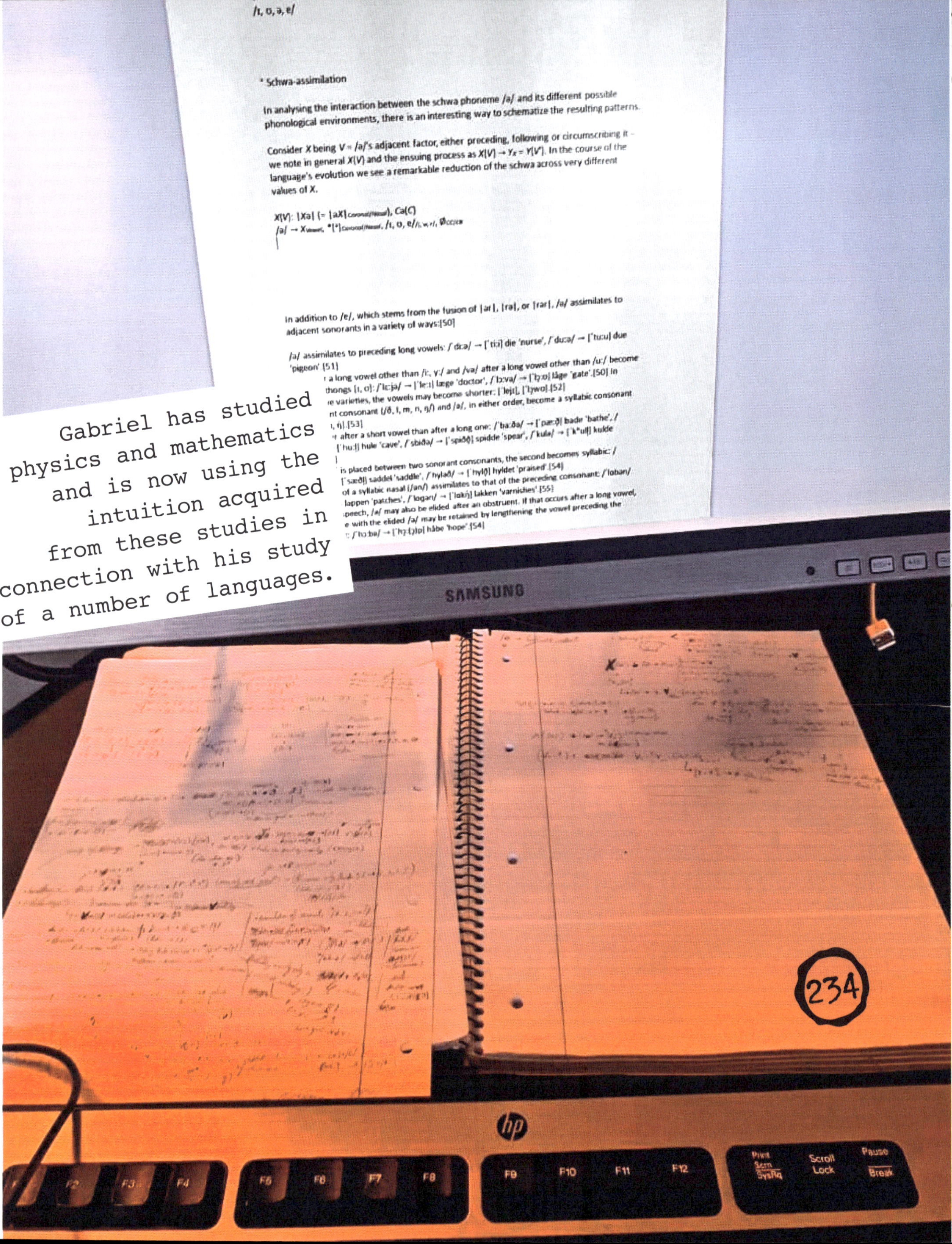

Gabriel has studied physics and mathematics and is now using the intuition acquired from these studies in connection with his study of a number of languages.

In *-rg*, /g/ in */rg/ disappeared probably by weakening and merging with /r/ due to and somewhere along the evolution of /r/ = [r]$_{\textit{Old Norse}}$ > [ʁ̞] - i.e. _/rg/ > *_[ʁ]/g/ > *_[ʁ̞] > _[ɐ̯]; e.g. *spurgte*, *spørgsmål*. Regarding *Vgl*, we have possibly the following evolution (notice the syllabification) with *V* = *i/u*: **i/ug.l_* > */ij:uv/.*l_* > [i:/u:].*l_*, again with [ij/uw] → [i:/u:]. Perhaps analogical processes occurred with other values of *V* (if such segments do exist). Examples are e.g. *fugl*, *kugle*, *ugle*, *igle*.

° Schwa-assimilation

In analysing the interaction between the schwa phoneme /ə/ and its different possible phonological environments, there is an interesting way to schematize the resulting patterns.

Consider *X* being *V* = /ə/'s adjacent factor, either preceding, following or circumscribing it - we note in general *X*{*V*} and the ensuing process as *X*{*V*} → Y_X = *Y*{*V'*}. In the course of the language's evolution we see a remarkable reduction of the schwa across very different values of *X*. Namely, *X* = *V*:, {$d_{/ð/}$, *l*, *n*}$_{Coronal}$, {*j*, *g*:$v_{/j:w/}$, $r_{/ɐ/}$}$_{Glide}$.

X{ə}: |*X*ə| (= |ə*X*|$_{Coronal,/r/}$), *C*ə(*C*)$_{Elision}$
/ə/ → *V*, |*X*|$_{κ}$ = /ɪ:ʊ, ɐ/$_{/j:w,\ ɐ/}$, $X_{Syllabification}$ (< *[ᵊ]$_{Coronal}$), Ø$_{CC/C\#}$

Let us exemplify each instance of this, case by case:

- Long vowels (*V*:/ə/ → *V*:*V*).

/ə/ assimilates to preceding long vowels as in the words *die*, *due* /ˈdi:ə, ˈdu:ə/ → [ˈti:i, ˈtu:u].

- Glides (|*X*ə| → |*j*, *g*:*v*, *r*|$_{κ}$)

The morphophonemic formulas |*X*ə|: |jə|, |gə|, |və|, |rə| = |(r)ər| are in fact phonemically realized as if they were a single element |*X*ə| = |*X*|, which becomes then the coda of the previous syllable, with the schwa syllable being effectively merged into it: _*V*.|*X*ə| → _*V*|*X*|$_{κ}$ = _*V*/ɪ:ʊ, ɐ/.

Clearly, the result is a diphthong, and it can be a long one if *V* = *V*: (see below), as in the following examples: e.g. *ture*(*r*), *veje*, *jage*, *have* /tu:ɐ, 'vɑjə, 'ja:jə, 'ha:və/ → [tsʰu:ɒ̽, ʋɑ̈ɪ, jɛ:ɪ, hɛ:ʊ]. As a side note, only |ər| = [ɒ̽] normally follows a consonant and obviously doesn't form a diphthong, e.g. *fatter* = /'fa.dɐ/ = ['fæ.tɒ̽].

When the vowel is already long, the diphthong becomes long - i.e. V:V': e.g. *læge*, *låge* /ˈlɛ:jə, ˈlɔ:və/ → [ˈle:ɪ, ˈlɔ̞:ʊ]; in innovative varieties, the vowels may become shorter: [ˈlejɪ], [ˈlɔ̞wʊ]. Notice words like *pige*, where we have /pijə/ → /pi:ə/ → [ˈpʰi:i] by virtue of the rule for long vowels above.

The stronger the sun, the more effective the solar cooker is.

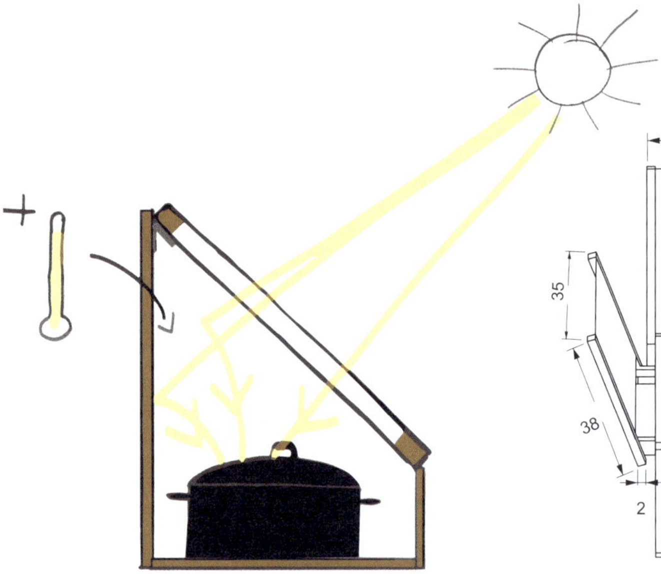

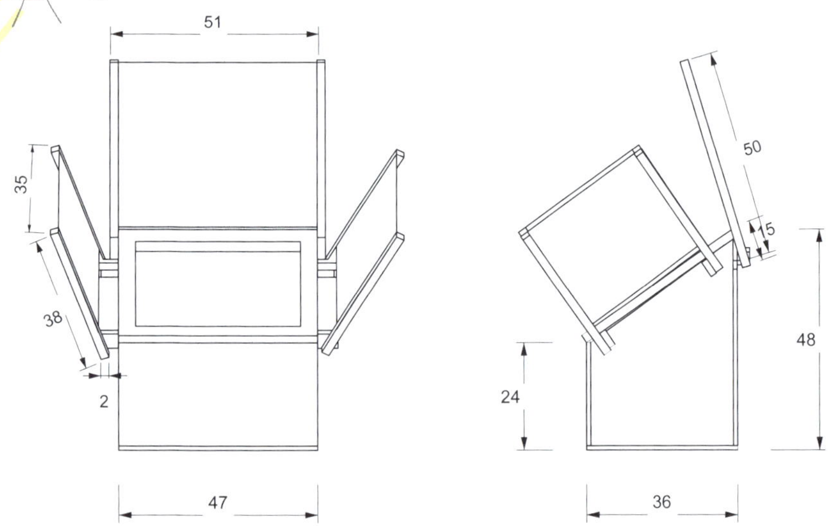

The solar cooker can be constructed from an old window, recycled wood and a reflecting material such as the inside of used potato-chips bags. It is light and mobile and can be used on both land and sea, requiring only energy from sun to work.

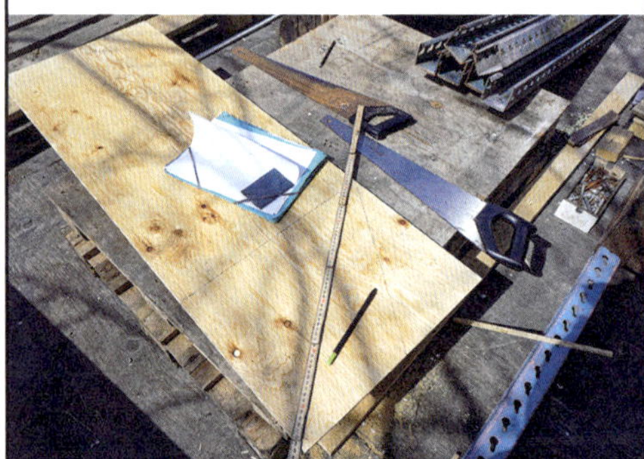

Solar cooker

At the age of 16, Kasper met a group of Germans at the Roskilde Festival They fell in love with both his music and his personality. In that sam year, he had survived a fall from the rooftop of a 4-story building resulting in serious injuries and chronic pain. At that time, he wa already living on the street after a challenging childhood. Music enable him to overcome these difficulties and maintain a positive attitude to life

Kasper Hedegreen

he Germans helped him perform at various music festivals in Germany and
hey set up the 'Kasper Support Foundation', which purchased a ship for
im to live on and paid harbor-rent in Copenhagen. After about a year,
owever, Kasper decided to move to the Pirate Harbor in Erdkehlgraven,
here no rent had to be paid, but the Germans remained close friends
nd a second family, with whom he still spends vacations each year.

The floating university in Berlin by Raumlabor is an offshore laboratory for collective experimental learning, knowledge transfer and the formation of trans-disciplinary networks aimed at challenging routines and habits of urban practices.
Roman Karrer – graphic mastermin

P.S. The authorities seem to believe the word 'university' is state property: this is why it has been crossed out here.

FLOATING UNIVERSITY BERLIN 3

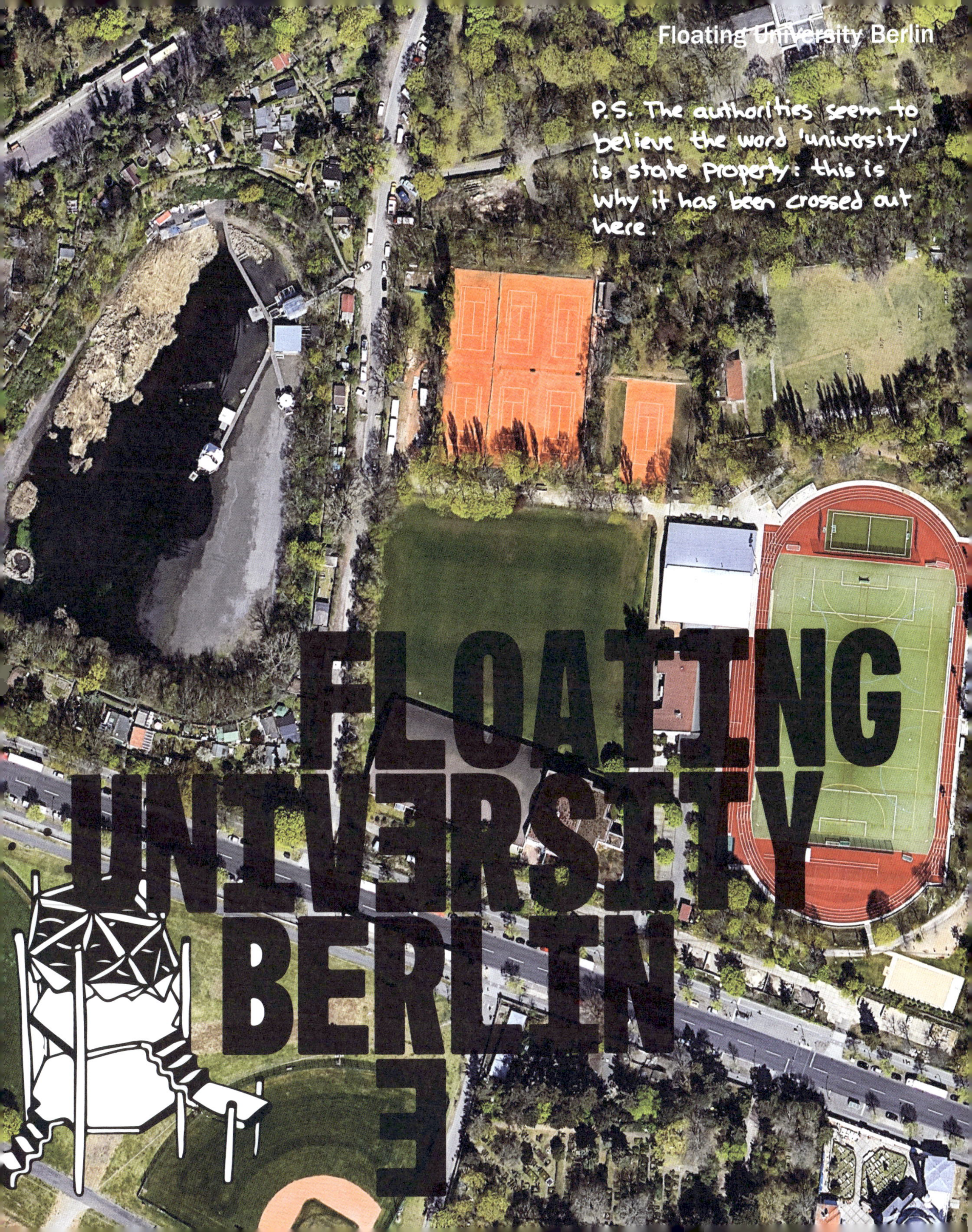

"THE 'SEA WOLF' IS A SELF-CREATED MYTHICAL BEING, HALF DRAGON, HALF PHOENIX. SCULPTED WITH AN AXE, CHISELS, AND HAND-SANDED. TREATED WITH LINSEED OIL AND FIRED SO AS TO RESIST THE WEATHER. OVER A PERIOD OF THREE MONTHS, AS THIS PIECE PROGRESSED, A GROUP OF YOUNG PEOPLE JOINED ME IN OPEN WORKSHOPS CENTERED ON PAINTING AND MOVEMENT."

"THE CREATION OF THE OVEN WAS ANOTHER GOOD EXAMPLE OF COLLABORATION, WITH BRICKS AND STONES COLLECTED BY THE GROUP AND ORGANIZED BY MARTHA AND PALI, WHO CAME AND HELD THE CONSTRUCTION WORKSHOP, LEAVING THE FINISHING PROCESS TO ME. PALI WOULD BRING HIS SON, AND THE AMBIENCE OF COOPERATION AND PLAY MADE IT CLEAR THAT THE PRESENCE OF THE FUTURE GAVE REASON TO THE PRESENT. AN ATTITUDE OF CARE PRECLUDES THE NEED FOR RULES."

THE MODERN CITY OF

development has increasingly been driven
who are acting on behalf of investors and
managers who are acting on behalf of po
shared and vested interest in stability and
parts – are constructed in an instant, afte
suddenly demolished. There are no trace
have been reduced to a bare minimum.
consumers enjoy a life of comfort and saf
of for them. Everything, including their de
planned and designed and is being careful
urban life. But isn't life exactly that which
is not regular and regulated, predete

THE SETTLED

In recent decades, urban
developers and other so-called consultants
anciers, and controlled by technicians and
cians and administrations, who all have a
redictability. These modern cities – or city
which they remain the same, until they are
f change and the potentials for surprises
e users or, perhaps, more precisely, the
, in which everything has been taken care
es and their behavior, has been carefully
monitored and managed. We call it modern
nnot be driven and controlled, that which
ined and predictable, stable and secure?

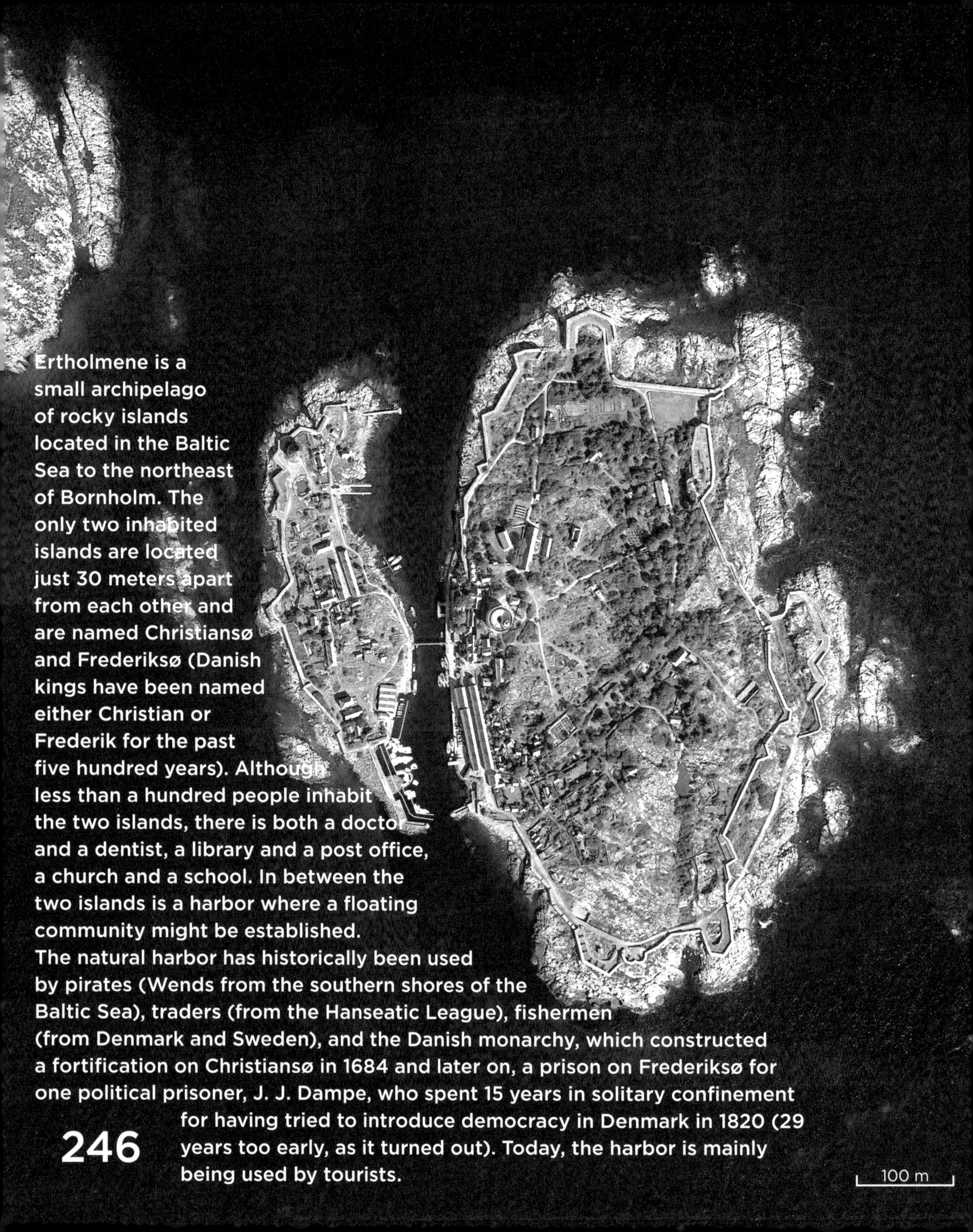

Ertholmene is a small archipelago of rocky islands located in the Baltic Sea to the northeast of Bornholm. The only two inhabited islands are located just 30 meters apart from each other and are named Christiansø and Frederiksø (Danish kings have been named either Christian or Frederik for the past five hundred years). Although less than a hundred people inhabit the two islands, there is both a doctor and a dentist, a library and a post office, a church and a school. In between the two islands is a harbor where a floating community might be established. The natural harbor has historically been used by pirates (Wends from the southern shores of the Baltic Sea), traders (from the Hanseatic League), fishermen (from Denmark and Sweden), and the Danish monarchy, which constructed a fortification on Christiansø in 1684 and later on, a prison on Frederiksø for one political prisoner, J. J. Dampe, who spent 15 years in solitary confinement for having tried to introduce democracy in Denmark in 1820 (29 years too early, as it turned out). Today, the harbor is mainly being used by tourists.

Ertholmene
Shipyard and workshop
Community kitchen
Composting bins
Kitchen garden
Meeting area
Kitchen garden
Composting bins
10 m

The Warao inhabit the Orinoco Delta. In this area of rivers and rivulets, wetlands and marshes, mangroves and swamp forests, the canoe is the only efficient means of transportation. They use both a large canoe, the bongo, and a smaller one. It is said that the children learn to paddle before they can walk.

The Warao construct their houses on stilts in the water, which made the conquistador, Alonso de Ojeda, name the country Venezuela (Little Venice). The houses have thatched roofs but usually no walls. In the center there is a fireplace of clay, around which people sleep in hammocks, and next to the house is typically an open platform for docking canoes. A single house may accommodate an extended family or several nuclear families and a group of houses forms a community.

The inaccessibility of the area provides the Warao with some protection but it also makes it hard for health workers to get there. Many children die from easily preventable diseases. Despite the isolation from other people, HIV has been spreading widely among the Warao over the past decade.

The Warao mostly eat fish but they also practice horticulture and find fruits and vegetables in the wild orchards of the delta.

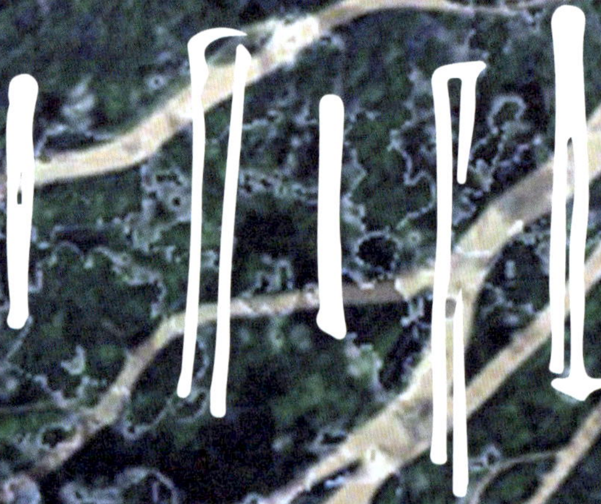

They make tightly hand-woven baskets of fibers wrapped around a coil of a palm branch and use natural vegetable dyes to color them.

They have shamans who
perform rain dances and songs.

"Allowable? – allowable.
Unallowable? – unallowable.
The Way comes about as we walk it … things however peculiar or incongruous, the Way interchanges them … to see as 'interchangeable' is to grasp …
It is this that is called 'Letting both alternatives proceed'."

Zhuangzi (4th century BC) in Graham, A. C. (1981)
Chuang-Tzu – The Inner Chapters. Indianapolis, Hackett, pp. 53-54.

MORE EXPERIMENTS,

understanding of humanity's common cha
that these challenges can – and will – be
breakthroughs, policy-making and managin
anything. In fact, we should just continue to
and we should just keep on voting for the sa
'green'). Unfortunately, everything seems to ir
only way to solve the challenges is by changi
and one Danish researchers stated that th
life is compatible with a climate-friendly age
calculations and dubious speculations. This
community, and many more experiments like
spiritual potentials and can serve to enh

NOT LESS

Over the past few decades, a greater

nges has emerged, together with a notion
olved through scientific and technological
We, the common people, don't have to do
y the same products (which are now 'green')
politicians (who, by now, have also become
cate that this notion is an illusion and that the
our way of life. Thus, in 2018, three hundred
ery notion that our – the well-off's – way of
a is based on false assumptions, misleading
why we need experiments like the floating
that can serve to stimulate our creative and
ce our awareness of nature and of each other.

MAKE WEEKEND

2) online check (..., ...), boat sell pick up. Jakke snull 3,000

Monday 18 may: COWI glays clean

ASK
Write deal experiences on songs

"They will see the glimpse of the Devil,
Reflection on your soul
(You will say no to free coke
Run away as fast as you can)
And the Devil whispers:
"It's your time to shine",
there's no way around in
you will charm them
with your delusional desires,
Because what you really
want of not from this world
(physical realm)

Song thoughts

• Cant you see I'm broke?
Get on here for free a...
Didn't even make a b...
That guy gave it to me fo...
(Cant you see it in my e...
(He told me a bring Luck)
(with my appearance

• old song thoughts:
I've been saying
to make a Deal
with the Devil
Just so I could say
My Last You cannot fool me
goodbye Love you fool all of them
(Coce and drugs I am
Love bitch!

Piracy lifestyle apparels
→ Grab my jacket, throw it
on the floor, Fine
Kept on growing
More, More, Mooore
I took your heart
As a magic spell
from the dead world...
(when I finish vn)
4 years ago you in...
That was covered in
layers of snow
You cant curse me to hell,
I choosed to fly over while I'm gone"
And let another one
keep my heart on fire
whisper lust words in his ear,
saying you made me feel
like I'm inlove again
Saving my "ass" from
drugdealers, pushers, small term
criminals. etc.

Parahting! crap

Paint my face with tears,
got locked into a
broken mirror in
28 July, LV inputs

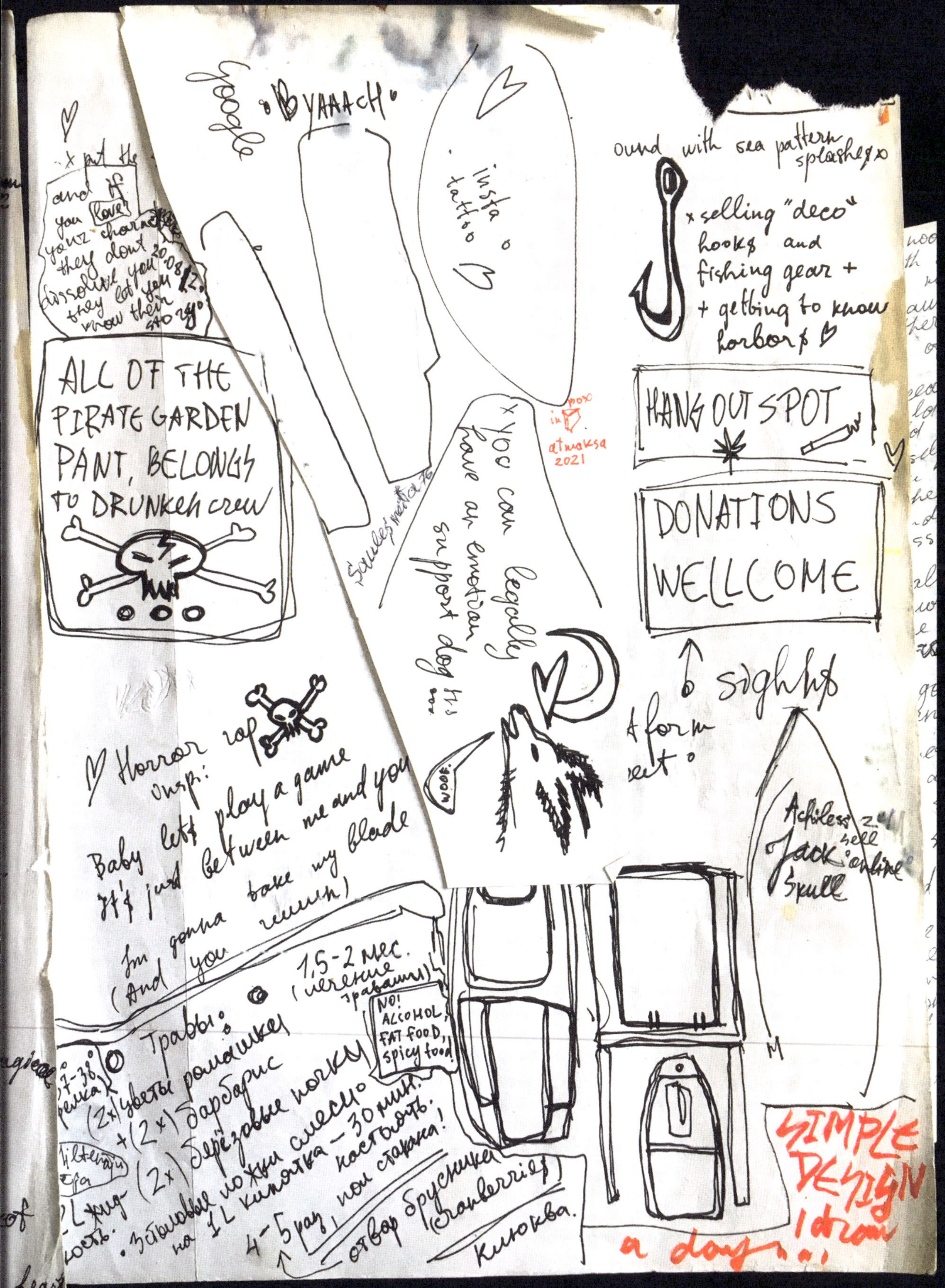
google
«Bуаaaсн»
insta
tattoo
and if you love your characters they dont dissolve you 20.08.12. they let you know their story
ALL OF THE PIRATE GARDEN PANT, BELONGS TO DRUNKEN CREW
ound with sea pattern splashes
× selling "deco" hooks and fishing gear + + getting to know harbors
HANG OUT SPOT
DONATIONS WELLCOME
× you can legally have an emotional support dog it's ...
atmaksa 2021
sights
Horror rap
Baby lets play a game
It's just between me and you
Im gonna take my blade
(And you ...)
1,5-2 мес. (лечение травами)
Травы:
NO! ALCOHOL, FAT FOOD, spicy food!
(2×) цветы ромашки
+ (2×) барбарис
(2×) берёзовые почки
Achiless Jack Skull
2. sell online
SIMPLE DESIGN
a day
1 draw
отвар брусники (cranberries)
клюква

INDEX